Statistical Analysis of DNA Sequence Data

STATISTICS: Textbooks and Monographs

A SERIES EDITED BY

Vol. 1: The Generalized Jackknife Statistic, *H. L. Gray and W. R. Schucany*
Vol. 2: Multivariate Analysis, *Anant M. Kshirsagar*
Vol. 3: Statistics and Society, *Walter T. Federer*
Vol. 4: Multivariate Analysis: A Selected and Abstracted Bibliography, 1957-1972, *Kocherlakota Subrahmaniam and Kathleen Subrahmaniam* (out of print)
Vol. 5: Design of Experiments: A Realistic Approach, *Virgil L. Anderson and Robert A. McLean*
Vol. 6: Statistical and Mathematical Aspects of Pollution Problems, *John W. Pratt*
Vol. 7: Introduction to Probability and Statistics (in two parts), Part I: Probability; Part II: Statistics, *Narayan C. Giri*
Vol. 8: Statistical Theory of the Analysis of Experimental Designs, *J. Ogawa*
Vol. 9: Statistical Techniques in Simulation (in two parts), *Jack P. C. Kleijnen*
Vol. 10: Data Quality Control and Editing, *Joseph I: Naus*
Vol. 11: Cost of Living Index Numbers: Practice, Precision, and Theory, *Kali S. Banerjee*
Vol. 12: Weighing Designs: For Chemistry, Medicine, Economics, Operations Research, Statistics, *Kali S. Banerjee*
Vol. 13: The Search for Oil: Some Statistical Methods and Techniques, *edited by D. B. Owen*
Vol. 14: Sample Size Choice: Charts for Experiments with Linear Models, *Robert E. Odeh and Martin Fox*
Vol. 15: Statistical Methods for Engineers and Scientists, *Robert M. Bethea, Benjamin S. Duran, and Thomas L. Boullion*
Vol. 16: Statistical Quality Control Methods, *Irving W. Burr*
Vol. 17: On the History of Statistics and Probability, *edited by D. B. Owen*
Vol. 18: Econometrics, *Peter Schmidt*
Vol. 19: Sufficient Statistics: Selected Contributions, *Vasant S. Huzurbazar (edited by Anant M. Kshirsagar)*
Vol. 20: Handbook of Statistical Distributions, *Jagdish K. Patel, C. H. Kapadia, and D. B. Owen*
Vol. 21: Case Studies in Sample Design, *A. C. Rosander*
Vol. 22: Pocket Book of Statistical Tables, *compiled by R. E. Odeh, D. B. Owen, Z. W. Birnbaum, and L. Fisher*
Vol. 23: The Information in Contingency Tables, *D. V. Gokhale and Solomon Kullback*

Vol. 24: Statistical Analysis of Reliability and Life-Testing Models: Theory and Methods, *Lee J. Bain*
Vol. 25: Elementary Statistical Quality Control, *Irving W. Burr*
Vol. 26: An Introduction to Probability and Statistics Using BASIC, *Richard A. Groeneveld*
Vol. 27: Basic Applied Statistics, *B. L. Raktoe and J. J. Hubert*
Vol. 28: A Primer in Probability, *Kathleen Subrahmaniam*
Vol. 29: Random Processes: A First Look, *R. Syski*
Vol. 30: Regression Methods: A Tool for Data Analysis, *Rudolf J. Freund and Paul D. Minton*
Vol. 31: Randomization Tests, *Eugene S. Edgington*
Vol. 32: Tables for Normal Tolerance Limits, Sampling Plans, and Screening, *Robert E. Odeh and D. B. Owen*
Vol. 33: Statistical Computing, *William J. Kennedy, Jr. and James E. Gentle*
Vol. 34: Regression Analysis and Its Application: A Data-Oriented Approach, *Richard F. Gunst and Robert L. Mason*
Vol. 35: Scientific Strategies to Save Your Life, *I. D. J. Bross*
Vol. 36: Statistics in the Pharmaceutical Industry, *edited by C. Ralph Buncher and Jia-Yeong Tsay*
Vol. 37: Sampling from a Finite Population, *J. Hájek*
Vol. 38: Statistical Modeling Techniques, *S. S. Shapiro*
Vol. 39: Statistical Theory and Inference in Research, *T. A. Bancroft and C.-P. Han*
Vol. 40: Handbook of the Normal Distribution, *Jagdish K. Patel and Campbell B. Read*
Vol. 41: Recent Advances in Regression Methods, *Hrishikesh D. Vinod and Aman Ullah*
Vol. 42: Acceptance Sampling in Quality Control, *Edward G. Schilling*
Vol. 43: The Randomized Clinical Trial and Therapeutic Decisions, *edited by Niels Tygstrup, John M. Lachin, and Erik Juhl*
Vol. 44: Regression Analysis of Survival Data in Cancer Chemotherapy, *Walter H. Carter, Jr., Galen L. Wampler, and Donald M. Stablein*
Vol. 45: A Course in Linear Models, *Anant M. Kshirsagar*
Vol. 46: Clinical Trials: Issues and Approaches, *edited by Stanley H. Shapiro and Thomas H. Louis*
Vol. 47: Statistical Analysis of DNA Sequence Data, *edited by B. S. Weir*

OTHER VOLUMES IN PREPARATION

Statistical Analysis of DNA Sequence Data

edited by

B. S. Weir

Department of Statistics
North Carolina State University
Raleigh, North Carolina

MARCEL DEKKER, INC. ***New York and Basel***

Library of Congress Cataloging in Publication Data

Main entry under title:

Statistical analysis of DNA sequence data.

(Statistics, textbooks and monographs ; v. 47)
Bibliography: p.
Includes index.
1. Deoxyribonucleic acid--Analysis--Addresses, essays, lectures. 2. Nucleotide sequence--Statistical methods--Addresses, essays, lectures. I. Weir, B. S. (Bruce S.), II. Title: Statistical analysis of DNA sequence data. III. Series.
QP624.S7 1983 574.87'3282'028 83-5137
ISBN 0-8247-7032-3

MARCEL DEKKER, INC.
270 Madison Avenue, New York, New York 10016

Current printing (last digit):
10 9 8 7 6 5 4 3 2 1

PRINTED IN THE UNITED STATES OF AMERICA

PREFACE

This book is intended to survey the rapidly growing field of statistical analysis of DNA sequence data. The authors are all engaged in such analyses, and several of them are also involved in the generation of DNA data. They have pointed to current problems in the interpretation of the new genetic information and have shown possible approaches to solving these problems. We all hope that the book will serve as a timely and convenient reference for molecular, population and evolutionary geneticists and will also serve to stimulate statisticians to become involved in one of the most exciting areas of modern science.

The disciplines of statistics and genetics are not strangers of course and there are many areas where the interaction of statisticians and geneticists have been very successful, as in plant and animal breeding or in evolutionary theory. The hybrid field of statistical genetics has also seen its share of controversy, dating from the split between the Mendelians and the Biometricians following the rediscovery of Mendel's Laws and continuing through the recent debate between the Neutralists and the Selectionists.

This most recent controversy was occasioned by the advent of electrophoresis for collecting protein data. This data posed several statistical problems when first employed by geneticists on a large scale some fifteen years ago. In contrast to studies with morphological characters, information became available routinely for several gene loci in the same individual. It became possible, and necessary, to accommodate interactions between different loci.

More challenging, however, was the need to account for the unexpectedly high level of protein variation discovered in natural populations. Statistical methods were needed to determine if observed patterns and levels of variation were consistent with the opposing neutral or selective theories of evolution. While the heat of debate has diminished in this area, there remain many unsolved problems for the analysis of protein data, but geneticists are collecting new data using recombinant DNA technology. New statistical problems are emerging.

In this book we consider two types of DNA data. The first type, on restriction fragments, arises when DNA is cleaved by restriction endonucleases. The data therefore consist of numbers and locations of bands, representing DNA fragments, on a gel. A statistical problem emerges immediately in finding the best way of determining the size of these fragments from the distances they migrate on electrophoretic gels. In the opening chapter of the book, Schaffer shows how some fairly basic statistical arguments offer great improvements over previous ad-hoc methods. The other data type we consider follows from further manipulation of DNA fragments. The fragments are sequenced to determine the constituent nucleotides in their physical order. The availability of sequences of thousands of nucleotides presents problems of managing the data even before statistical analyses are performed. The collection and storage of sequence data are very much dependent upon computers, and Gingeras has surveyed the crucial role that computers play in modern genetics. He lists available programs and indicates future needs.

While the statistical analysis of DNA data is still a young field, and serious controversies have not become apparent, there are already instances of alternative, and seemingly quite different, methods being proposed for interpreting data. In his chapter, Ewens reminds us of the absolute dependence of statistical results on the underlying models assumed. His treatment has relevance for all of applied statistics, but he considers restriction fragment data in particular and shows the relationships between apparently different analyses. Ewens' theme of exploring different models, and accounting for sources of variation in resulting parameter estimates, is continued by Kaplan. He considers DNA sequence and restriction fragment data and concentrates on measures of evolutionary distance and of variation within populations. When DNA sequencing first became possible, it was customary to prepare a sequence from a single representative of a species so that resulting comparisons between species were confounded by variation within species. Improvements in technology have now allowed larger samples to be taken, and a new range of statistical methods can be used. In their chapter, Brown and Clegg illustrate some methods by referring to replicate sequences collected for maize knob heterochromatin from within the same individual. Their work casts a new light on the consideration of replication provided by several loci as opposed to several populations.

If there is a present controversy in current statistical genetics, it arises in the area of reconstructing evolutionary history. Felsenstein describes the problems of inferring evolutionary trees from DNA sequences and shows how little use has yet been made of statistics. We hope that this book does succeed in attracting statisticians to these problems, and thereby disprove Felsenstein's gloomy prediction on the likelihood of progress in the near future. The same field of phylogeny reconstruction is discussed by Templeton who focuses on the phenomenon of convergent evolution. He points out that present sequences that are the same at particular sites may have arrived at that state following a series of changes from a single ancestral sequence, and need not have remained unchanged since that ancestral time.

One of the great hopes held for DNA sequence data, apart from general genetic and evolutionary applications as discussed in the first seven chapters, is that they will lead to advances in medicine. Restriction fragment sites are expected to be valuable genetic markers. They are relatively easy to assay and it is hoped that sites can be located close enough to disease loci to allow segregation analysis and perhaps diagnosis of genetic diseases. The efforts at the University of Utah to map the human genome by a suitable number of restriction site markers is described in the chapter by Bishop, Cannings, Skolnick and Williamson. The book closes with a mathematical treatment by Clegg and Asmussen on just how diagnoses can be aided by these markers, and how the likelihood of success depends upon a range of population parameters.

Space has prevented this book being more than a brief guide to a new application of statistics. The whole area of amino acid sequence analysis was not treated, for example, while the difficult statistical problems raised by considering several restriction sites or nucleotide sequences jointly were barely mentioned. I hope that this book will be of use, however, since it does survey a great deal of current activity and refers to a major portion of the current literature. The quality of the book is a tribute to the efforts of the thirteen authors, and it has been a pleasure for me to work with them. I am very grateful to them for agreeing to contribute to the book and for the care with which they not only prepared and revised their own chapters but also reviewed at least two other chapters. My particular thanks in the reviewing process go to Henry Schaffer.

The impetus for this book came from Dr. Maurits Dekker, and its appearance is a tribute to the professionalism of many people at Marcel Dekker, Inc. These pages are also testimony to the superb typing of Selma McEntire.

B. S. Weir

CONTENTS

CONTRIBUTORS

Marjorie A. Asmussen *Departments of Molecular and Population Genetics, and Mathematics, University of Georgia, Athens, Georgia*

D. Timothy Bishop *Department of Human Genetics, University of Utah, Salt Lake City, Utah*

A. H. D. Brown *Division of Plant Industry, CSIRO, Canberra, ACT, Australia*

Christopher Cannings *Department of Probability and Statistics, University of Sheffield, Sheffield, England*

Michael T. Clegg *Departments of Molecular and Population Genetics, and Botany, University of Georgia, Athens, Georgia*

W. J. Ewens *Department of Biology, University of Pennsylvania, Philadelphia, Pennsylvania* and *Department of Mathematics, Monash University, Clayton, Victoria, Australia*

Joseph Felsenstein *Department of Genetics, University of Washington, Seattle, Washington*

Thomas R. Gingeras *Cold Spring Harbor Laboratory, Cold Spring Harbor, New York*

Norman Kaplan *Biometry and Risk Assessment Program, NIEHS, Research Triangle Park, North Carolina*

Henry E. Schaffer *Department of Genetics, North Carolina State University, Raleigh, North Carolina*

Mark Skolnick *Department of Human Genetics, University of Utah, Salt Lake City, Utah*

Alan R. Templeton *Department of Biology, Washington University, St. Louis, Missouri*

John A. Williamson *Department of Mathematics, University of Colorado, Boulder, Colorado*

Statistical Analysis of DNA Sequence Data

Chapter 1

DETERMINATION OF DNA FRAGMENT SIZE FROM GEL ELECTROPHORESIS MOBILITY

Henry E. Schaffer

Department of Genetics, North Carolina State University, Raleigh, North Carolina

I. INTRODUCTION

When the DNA of an organism is being extracted and purified for study, it breaks into pieces. It is common to cut it into shorter lengths of manageable size, and during the early studies of DNA this was done by mechanical shearing. This could produce lengths of relatively uniform size, but breakage was not related to the DNA base sequence so that the information obtained from these lengths was limited. With the discovery and widespread use of restriction endonucleases, it became possible to cut the DNA in precisely repeatable locations at known recognition sequences. These restriction fragments can then be separated and replicated by cloning them in bacterial plasmids (Bolivar, 1979; Morrow, 1979). This can give a large quantity of a particular fragment to facilitate further analysis. Wu (1979) has given a survey of the methods used in manipulating and studying DNA molecules. This chapter is concerned with a methodology for inferring the sizes of the fragments produced by restriction endonucleases.

II. PRODUCTION AND ELECTROPHORESIS OF DNA FRAGMENTS

At the start of many analyses there is a sample of DNA which contains many copies of the same sequence. It may be desired to partially characterize this DNA molecule, such as by restriction mapping to produce a physical map, or to study it more fully by determining the complete base sequence. In either case it is usually necessary to break it into smaller fragments. Complete determination of the base sequence has become much more feasible recently, but it is still a time consuming process. Since the sequencing techniques currently in use work best with short DNA fragments (of length at most several hundred base pairs), longer lengths of DNA must be cut up and separated as an intermediate step in the sequencing process.

The sample of DNA is treated with one or more restriction endonucleases that cut it at specific recognition sequences, ensuring that each copy of the DNA sequence is cut at exactly the same location(s). Many restriction enzymes are known, and their recognition sequences have been cataloged (Roberts, 1982). We will assume throughout that a DNA sample contains copies of only one sequence prior to treatment with restriction enzymes.

When DNA has been treated with a restriction enzyme, it is described as having been "restricted", and since each long strand of DNA has been cut at exactly the same places, the sample then has many copies of some number of smaller DNA fragments ("restriction fragments"), which must be separated and characterized. Separation is most commonly done by gel electrophoresis. The gel is a solid matrix, typically of agarose or polyacrylamide, permeated with a liquid phase buffer. The gel is placed in a direct current electrical field (see Figure 1a) and a sample of DNA is placed near one end. Because the DNA in the gel is a negatively charged molecule, it will migrate towards the positive pole of the electrical field (Southern, 1979a).

It has been observed that the different fragments in a sample often have different mobilities, that is, they migrate at different

a.

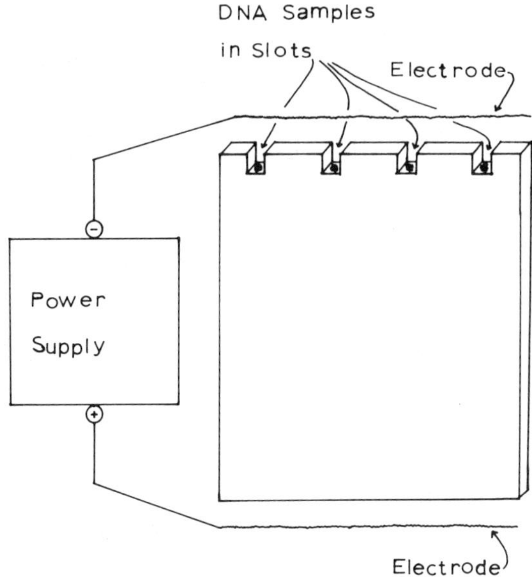

b.

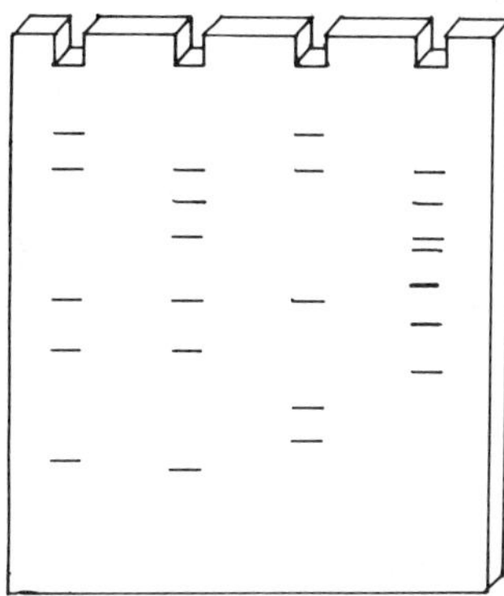

Figure 1. (a) Gel with DNA samples in slots at beginning of electrophoresis procedure. In practice, there may be ten or more slots. (b) The same gel after detection of the DNA at the conclusion of the electrophoresis run.

speeds relative to each other. This can be detected by removing the gel from the electrophoretic apparatus and staining the gel with ethidium bromide. The combination of DNA and ethidium bromide fluoresces strongly, enabling the locations of the DNA fragments to be seen when the gel is illuminated with ultraviolet radiation. A more sensitive technique, which can be used with smaller quantities of DNA, involves using radioactively labeled DNA and then placing a sheet of X-ray film next to the gel. The radioactivity exposes the film wherever there is DNA and the locations of DNA can be seen on the developed film. For reasons of economy, many samples of DNA can be run at the same time on one gel, and a typical situation is shown in Figure 1b.

At this point there are one or more restriction fragments per band on the gel, since different fragments may have the same mobilities. It is usually desired to describe the fragments in terms of their sizes, and this is done by relating mobilities and sizes.

The size of a double-stranded (duplex) DNA molecule can be expressed in several ways. The molecular weight is expressed in daltons, and the length can be expressed in base pairs (bp). While the conversion between molecular weight and base pairs depends on the exact base composition it is common to use the approximation (Davis *et al.*, 1980) that 1 bp is about 660 daltons. The physical length of a DNA molecule depends on its conformation, which in turn is affected by many factors. While theoretical calculations indicate a length of 3.4 Å per bp, electron micrograph estimates are quite variable and, for example, Schleif and Hirsh (1980) show a length of approximately 2.7 Å per bp. Because of the greater variability in length as shown in an electron micrograph, it is customary to express length in terms of another, standard, DNA molecule which has been included in the micrograph, rather than in terms of an absolute length.

Since these different measures of size are proportional to each other, any of them can be used, and in most analyses it makes no difference which is used as long as there is consistency. As sequencing has become more common, measurement in base pairs and in molecular weight is being used. Measurement of fragment size in terms of physical length is seldom done outside of applications of electron microscopy. In this discussion L_i will be used to represent the size of the *ith* fragment. (Other authors have used symbols such as M, MW, N and S.)

A wide range of DNA fragment lengths can be separated and characterized by means of gel electrophoresis. Generally, agarose gels are used for the longer fragments and acrylamide gels for the shorter ones, with higher concentrations of agarose and acrylamide being used for shorter fragments. Agarose concentrations from 0.1% to 2.5% have been used for DNA molecules of 880,000 to 150 bp. Acrylamide concentrations from 3% to 20% have been used for shorter molecules of 200 to 1 or 2 bp (Bostian *et al.*, 1979; Yang *et al.*, 1979).

Mobility of a fragment in a given gel system is defined as the distance traveled in the gel divided by the product of the electrical field strength (measured in volts per unit distance) and the time (in seconds). In practice, however, it is most common to work with the distance traveled in the gel, generally as measured on a photograph or autoradiograph of the gel. In a given gel system this distance is proportional to mobility. Because of the gel-to-gel differences, analyses are done only within a gel and distances are used as relative mobilities. In this discussion m_i will be used to represent the mobility, or distance traveled, of the *ith* fragment. (Other authors have used such symbols as d, M, U, μ or x.) Since the measurement made for mobility is a migration distance, the two terms, mobility and migration, will be used interchangeably in this chapter, as is usual in the literature.

III. MODELS AND ANALYSES

It has long been known that longer fragments have lower mobility in gel electrophoresis, and it has been customary to plot mobility against the logarithm of size. A number of (standard) fragments of known sizes are run on the gel to establish a standard graph, and then the mobilities of the fragments of unknown size on that gel are entered on the graph to estimate their sizes.

For some time it was believed that there was a linear relationship between mobility and log size, and that the failure of the standards to give a perfectly straight line plot was due to imperfect knowledge of the true lengths of the standards. However, it soon became apparent that the relationship was actually curved and concave upwards (Figure 2), and it became the practice to use different methods of interpolation in the straight and curved portions of the curve. Linear regression was often used for the straight portions, and "eyeball" interpolation used for the curved region.

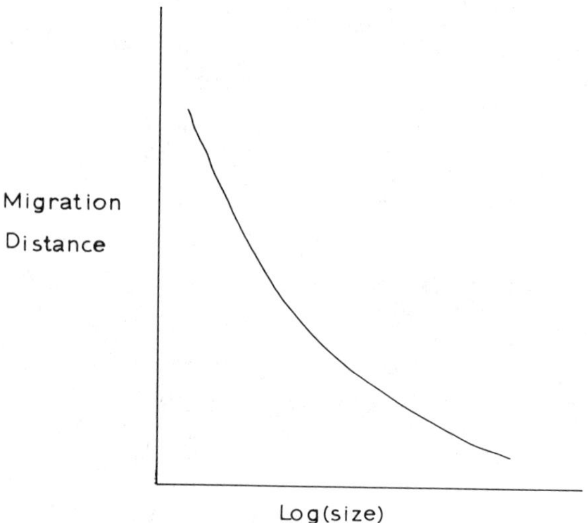

Figure 2. Shape of slightly curved line when migration is plotted against the logarithm of size on semi-log paper. The axes are sometimes reversed in the literature, with migration distance on the ordinate instead of the abscissa.

A linear relationship between mobility and the logarithm of size can be written as

$$m_i = k - c \log L_i$$

where k and c are positive constants. Empirically, it does fit mobility data reasonably well when only a limited range of sizes are included. This has led to the practice of plotting data from gels on semi-log paper and drawing a straight line through the points representing the standard fragments. Such a plot can be seen in Fangman (1978) and is referred to as approximating a straight line, although the curvature is easily noticeable. Daniels *et al.* (1980) used linear regression for the straight portions and a more subjective interpolation for the curved portion.

The introduction of a curvature can start with logarithmic or exponential polynomial relationships. Parker *et al.* (1977) used

$$L_i = \exp\,(a_0 + a_1 m_i + a_2 m_i^2 + a_3 m_i^3)$$

They mention that they explored various functions before settling on this one, and that they found the linear logarithmic function to be less satisfactory. A cubic polynomial fits nearly as well as this cubic exponential. The method of estimation used by Parker *et al.* (1977) did not involve the use of standards, but instead used other available information about the fragments being studied. The fragments all came from closed circular DNAs which were first cut once to produce linear DNA molecules. For example, they have used the restriction enzyme *Hind* III, with six recognition sites on the circular DNA of SV40. Use was then made of ethidium bromide, which intercalates itself within a DNA molecule and inhibits the restriction reaction. Since ethidium bromide intercalates more readily in linear than in supercoiled DNA, the reaction of the *Hind* III enzyme on SV40 was generally inhibited after causing the first cleavage of each molecule. A mixture of six full length linear molecules was thereby produced which were circularly permuted arrangements of each other. Then, in the absence of ethidium bromide, they were treated with another restriction enzyme which had only a single recognition sequence in this molecule. For example, *Hpa* II was used to cut each circularly permuted linear molecule into exactly two pieces. These could be recognized as pairs, the shortest with the longest and so on, from the fact that the pairs summed to a constant size. Additional information is available from the migration of the uncut linear molecule, and from a comparison of the six fragments produced by a complete *Hind* III digestion with the seven produced by dual digestion by *Hind* III and *Hpa* II. From the migration distances of all these fragments, with the total molecular size normalized to unity an over-determined set of equations resulted. These were solved by iterative least squares.

It is difficult to assess the adequacy of fit of the cubic exponential for the relationship between mobility and size. With four parameters being estimated, it has the potential of fitting the data quite well, but Parker *et al*. (1977) used it in a context with other information which is not generally available.

Duggelby *et al*. (1981) used a function with one less parameter:

$$m_i = a_0 + a_1 \log L_i + a_2 (\log L_i)^2$$

which has a polynomial in log size rather than in mobility. This function can be fitted by least squares regression of mobilities on the logarithms of the known sizes of standard fragments. Duggelby *et al*. illustrated their method by analyzing a mixture of restriction fragments of the plasmid pBR322. The sequence of this plasmid is known (Sutcliffe, 1979), so that the exact length of the fragments can be determined and compared to the estimates produced by an analysis of gel electrophoretic data. The example they gave is of fourteen *Hae* III fragments, varying in length from 18 to 267 bp. The agreement between the known lengths and those predicted from the best fitting quadratic is excellent.

Duggelby *et al*. (1981) evaluate the quality of the fit of their model by predicting in a reverse manner. They enter the standard curve, not with the mobility but with the known size, and then find the mobility which would have predicted this size. This can be compared with the actual mobility and the difference evaluated in terms of the difficulties of measurement and of the irregularities in the gel. In their example the standard deviation of these differences (labeled "STD ERR OF FIT" in their program) was 0.28 mm, which was not much larger than the errors made in measuring the mobilities. Duggelby *et al.* mention that the fit of the quadratic was less satisfactory when the standards varied greatly in size.

Another method of estimation of size from mobility data was developed by Schaffer and Sederoff (1981), based on a model of reciprocal mobility of Southern (1979a,b,c). The function used was:

$$(m_i - m_0)\ (L_i - L_0) = c$$

where 0 subscripts refer to "offset" values. The three parameters c, m_0, L_0 were estimated by least squares from standard fragment data. The parameter m can be interpreted as correcting the mobilities to an "apparent" origin, instead of the actual origin of the gel. Since conditions at the start of the electrophoresis run are often different between runs, this parameter can be said to have an interpretable meaning. As pointed out by Schaffer and Sederoff, the interpretation of L_0, however, is not clear. With such a nonlinear model there are a variety of approaches which can be used for fitting. Schaffer and Sederoff rewrote the model with c_i instead of c for the *ith* fragment, and then used a least squares approach to estimate m_0 and L_0 by minimizing the corrected sum of squares of the c_i. They presented a closed form solution to the normal equations.

The example given by Schaffer and Sederoff (1981) used ten standard fragments ranging from 843 to 5241 bp in length. These were from the bacteriophage *lambda* which had not been sequenced at that time. The fragment lengths were not known exactly but were used to predict the lengths of five fragments of pBR322 for which the sequence is known. Schaffer and Sederoff evaluate the quality of fit of the model by comparing the known length of a fragment with the length predicted by using its mobility along with the estimates of m_0, L_0 and c.

All of their predictions for the *lambda* and pBR322 fragments agreed with the known lengths within 1%. This discrepancy is comparable to the errors in measuring the locations of the DNA fragments on the gel.

Analyses of short chains of DNA (oligonucleotides) have shown that mobility depends on the base composition (Wu *et al.*, 1974). Electrophoresis in these experiments was done on cellulose acetate, so the results may not be directly applicable to the agarose and acrylamide experiments. Also, in longer chains of DNA the base composition should be less variable than with oligonucleotides. These factors may explain why Schaffer and Sederoff (1981) saw no evidence of sequence dependent migration.

IV. DISCUSSION

The quadratic and reciprocal models each involve the estimation of three parameters and so can be compared directly. Duggelby *et al.* (1981) and Schaffer and Sederoff (1981) both include computer programs for closed form estimation of the parameters of the models. Both methods are implemented using calculations well within the capacity of a very small computer, or even a programmable calculator.

Both methods fit a curvilinear relation between size and mobility, but they do so with different functional forms. The two methods can be compared by using them to analyse the same data. Table 1 gives the results of using both analyses on the data discussed by Duggelby *et al.* (1981). The analysis of these authors gave the (slightly) better fit but the two methods agree to within a fraction of a base pair for thirteen of the standards. There was a difference of 1.7 bp in the two predictions of the 267 bp length of the fourteenth standard. Agreement between the two methods is thus very close for that data set.

Table 1. Analysis of data of Duggelby *et al.* (1981).

Length of Standard Fragment	Migration Distance	Predicted Lengths Using Methods of	
		Duggelby *et al.*	Schaffer & Sederoff
267	5.9	266.5	264.8
234	7.0	237.2	237.0
213	8.1	213.9	214.0
192	9.2	194.8	194.8
184	10.0	182.8	182.7
124	16.1	122.2	121.5
104	19.2	103.1	102.6
89	22.5	87.5	87.3
80	24.6	79.4	79.3
64	29.5	64.3	64.6
57	32.1	57.9	58.4
51	35.0	51.8	52.5
21	64.0	20.8	20.8
18	69.2	18.1	17.8

The two methods have also been applied to the data discussed by Schaffer and Sederoff (1981), with a summary shown in Table 2. The agreement here is not as close, with the method of Duggelby *et al.* giving a noticeably poorer fit, particularly for the longer standard fragments. It might appear then that the reciprocal mobility model has a somewhat wider range of applicability than does the quadratic regression. Of course, it is not safe to make such a conclusion on the basis of only two sets of data, especially since the two methods do not disagree greatly and there are many other factors which affect migration and which could have been different in the experiments leading to the two sets of data.

While the emphasis in this discussion has been on the mathematical modeling and statistical analysis of mobility, these are probably not the limiting factors at present. The methods of Parker *et al.* (1977), Duggelby *et al.* (1981) and Schaffer and

Sederoff (1981) have all found that the measurement of migration distance of a molecule introduces an error of the same magnitude as the total lack of fit of the model to data.

Table 2. Analysis of data of Schaffer and Sederoff (1981).

Length of Standard Fragment	Migration Distance	Predicted Lengths Using Methods of	
		Schaffer & Sederoff	Duggelby, *et al.*
5241	3.03	5245.0	5186.6
5070	3.11	5046.4	5015.4
4257	3.46	4280.1	4322.3
3510	3.89	3519.4	3584.0
1995	5.17	1980.0	1982.7
1880	5.30	1863.9	1860.9
1584	5.63	1592.7	1579.2
1350	5.95	1357.6	1340.7
941	6.60	948.4	946.5
843	6.80	837.8	846.3
4362*	3.40	4400.8	4434.9
3233*	4.07	3246.8	3308.4
2958*	4.28	2956.6	3009.8
1404*	5.89	1399.7	1383.0
1129*	6.29	1133.3	1120.6

*Not used in constructing the standard curve.

This error of measurement refers to the uncertainty of the location of the center of the band of DNA on the gel, or on the photograph or autoradiograph of the gel. Gels often exhibit curvature of bands in a channel, tailing of bands and asymmetry of bands. In addition, it is not always clear where the origin of migration is: the DNA is inserted into a slot in the gel, but it is often mixed with some other substance and it may be driven into the gel by a higher voltage applied for a short time. Johnson *et al.* (1980) showed that an increased mass of DNA in a channel both increases the mobility and distorts the bands, making them more asymmetrical. They demonstrated that the mobility of one fragment can be affected by the quantity of another fragment included in the same sample. Mobility also depends on the gel concentration (Johnson and Grossman, 1977), so that any local variation in gel concentration will also contribute to the variability in migration.

We measure the location of bands on a photograph of a gel with a dial caliper micrometer. While we find this to be repeatable, both within and between observers, the resolution appears to limit substantial improvements in prediction capability. Improvements in the photographic method and use of a densitometer may yield improvements in resolution. The use of gradient gel systems, as suggested by Parker *et al.* (1977), would further increase resolution, but at the cost of introducing a nonuniformity in the gels which would make modeling and prediction of size from mobility data much more difficult.

A large DNA molecule is often studied by analyzing the fragments generated by several different restriction enzymes, and the lengths of the different fragments can be used to construct a restriction map of the whole genome. This is easiest when the lengths are known very accurately, but Schroeder and Blattner (1978) have presented a least squares method for producing a restriction map even when discrepancies are present in fragment lengths. This method was used by Daniels (1980) and is referred to by Gingeras in his chapter in this volume. The next analysis to proceed after completion of the restriction map usually involves sequencing the individual fragments. Sequence data are very extensive, and their handling and analysis need to be computerized, as discussed by Gingeras.

V. CONCLUSIONS

The quadratic model of Duggelby *et al.* (1981) and the reciprocal model of Southern (1979a,b,c) (Schaffer and Sederoff, 1981) are both simple in structure and both fit the gel migration data quite well. They both lead to analyses which are easy to perform. The reciprocal model is closer to being a model of migration in terms of interpretable parameters, and it may fit over a wider range of nucleotide sizes. However, comparisons based on data in the literature indicate that the performance of each model is limited by extraneous factors acting on DNA migration and by limits of accuracy in the measurement of migration distance.

To distinguish between models of mobility and to allow the effect (or lack of effect) of base sequence on mobility to be determined, it will be necessary to produce increased accuracy in the measurement of migration distances and to achieve better control of the gel system. With improved techniques it should be possible to determine fragment size within a small fraction of one percent. This would increase the speed and accuracy with which restriction maps may be constructed and would increase the value of the already widespread use of gel electrophoresis.

ACKNOWLEDGMENTS

This is Paper Number 8505 of the Journal Series of the North Carolina Agricultural Research Service, Raleigh, North Carolina. This research was supported in part by Grant Number GM 11546 from the National Institutes of Health.

Chapter 2

COMPUTERS AND DNA SEQUENCES: A NATURAL COMBINATION

Thomas R. Gingeras

Cold Spring Harbor Laboratory,
Cold Spring Harbor, New York

I. INTRODUCTION

The growth of interest in the use of computers to store and analyze nucleic acid sequence data has paralleled the explosive increase in the accumulation of DNA sequences. This parallel development is a natural result of the need to manage large quantities of detailed information. During the early phase of computer program development in this area there was little effort to capitalize upon the experiences of workers in other fields who have faced similar challenges in data manipulation. This oversight is perhaps understandable since the required manipulations were of a relatively simple variety: direct searches, best-fit analyses, data formatting, and so on. Such tasks were often separated, one from another, and the results combined in some manual fashion.

Molecular biologists have now become more ambitious and their interests are more complex. Two important pursuits have become the focus of more recent computer programs. One goal is the correlation of sequence data with results obtained from other methodologies. Examples of information from other disciplines include electron microscopic measurements, genetic mapping information, restriction enzyme sites and areas of known biological importance (splice points, promoter sites, polyadenylation sites, and so on). The second goal involves the ability to detect, within large bodies of sequence data, regions which contain specific biological signals. Since it seems that many such signals follow a consensus rather than a definitive format, learning algorithms offer one approach to the location of regions affecting biological control. A survey of the currently available computer programs used in the analysis of nucleic sequence data shows a consistent interest in these two topics. Consequently, this review is divided into three parts. The first presents a brief survey of the programs developed to perform straightforward analytical tasks. The second describes methods that are being used to correlate sequence and nonsequence data. The third contains an example of how computers could be used to extract the more subtle information that is inherent in structures of nucleotide sequences.

II. AVAILABLE COMPUTER PROGRAMS

A. Early Efforts

Recently, two very large blocks of sequence data have been determined. These are the nucleotide sequences of the bacteriophages *lambda* (48,000 bp, Sanger *et al.*, unpublished) and *T7* (39,000 bp, Dunn *et al.*, unpublished). The assembly of such large blocks of sequence data requires that a great deal of detailed information be assembled into ever-growing sequence blocks. The nature of the primary data from which nucleic acid sequences are assembled is shown in Figure 1. The information derived from many such autoradiographs of sequencing gels is necessary to accumulate blocks of sequences the size of the *lambda* or *T7* genomes. It became immediately obvious, during the earliest stages of most large sequencing projects, that computers would be essential to collect, manipulate and analyze the newly generated sequence data.

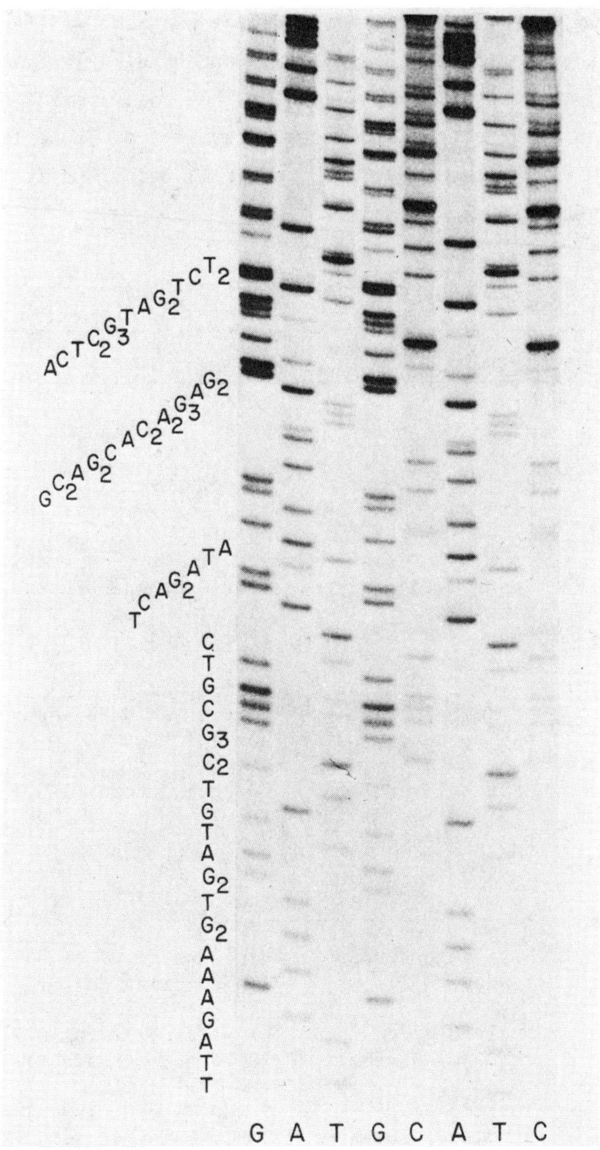

Figure 1. An autoradiograph of a sequencing gel. The sequence is read from the bottom to the top (5' to 3') of the figure by noting the channel containing the next highest band. Duplication of channels allows for every combination of nucleotides to be next to one another for comparison purposes. The sequence derived from this experiment is detailed on the side of the autoradiograph.

Consequently, the first types of computer programs written were those which helped in the organization and handling of DNA sequencing projects. Programs written during that early period of rapid accumulation of sequence data were the subject of an earlier review (Gingeras and Roberts, 1980). The programs described in that early survey are summarized in Table 1.

Table 1. Original computer programs for DNA sequences.

TITLE	FUNCTION	SOURCE	LANGUAGE
I. Presequencing Preparations			
GA1	Restriction enzyme site mapping.	Friedland *et al.* (Stanford U.) (Stefik, 1978)	SAIL
LEAST SQUARES METHOD	Restriction enzyme site mapping	Schroeder and Blattner (1978) (U. of Wisconsin)	FORTRAN
REVTRANS	Reverse translation of amino acids into DNA sequence.	Blumenthal *et al.* (1982) (Cold Spring Harbor)	FORTRAN
noname	Reverse translation of amino acids into DNA sequence.	Queen and Korn (1980) (NIH)	PL/I
II. Collection and Assembly of Sequences			
READ	Semi-automated gel reader.	Gingeras *et al.* (1982) (Cold Spring Harbor)	FORTRAN
ASSEMBLER	Nucleic acid sequence assembly	Gingeras *et al.* (1979) (Cold Spring Harbor)	FORTRAN
OVERLAP-MELD	Nucleic acid sequence assembly.	Staden (1979, 1980b) (MRC, Cambridge)	FORTRAN
III. Analysis of Sequences			
noname	Printing, editing, storage, manipulation.	Queen and Korn (1980) (Stanford U.)	PL/I

Table 1 (continued)

TITLE	FUNCTION	SOURCE	LANGUAGE
noname	Printing, editing, storage, manipulation.	Brutlag *et al.* (1982) (Stanford U.)	SAIL
noname	Search routines, mapping, translation.	Staden (1977) (MRC, Cambridge)	FORTRAN
MONITOR	Restriction enzyme recognition site.	Gingeras *et al.* (1978) (Cold Spring Harbor) (also Fuchs *et al.*, 1978) (U. of Wisconsin)	FORTRAN
T-RNA	tRNA gene prediction.	Staden (1979) (MRC, Cambridge)	FORTRAN
noname	Secondary structure prediction	Pipas and McMahon (1978) (U. of Syracuse)	FORTRAN
noname	Secondary structure prediction	Studnicka *et al.* (1978) (UCLA)	APL
noname	Three-dimensional modeling.	Feldmann (1978) (NIH)	FORTRAN
noname	Three-dimensional modeling	Bina *et al.* (1980) (NIH)	BLISS

These programs were concerned with automated methods to assist in pre-sequencing preparations, such as in the production of restriction enzyme maps (Schroeder and Blattner, 1978; Stefik, 1978) or in the reverse translation of amino acid sequences into putative nucleotide sequences (Korn *et al.*, 1977; Queen and Korn, 1980; Blumenthal *et al.*, 1982). Other aspects of sequencing projects such as the collection, assembly and analysis of nucleic acid sequences were also covered by these programs (Korn *et al.*, 1977; McCallum and Smith, 1977; Staden, 1977, 1978, 1980b; Fuchs *et al.*, 1978; Gingeras *et al.*, 1978; Studnicka *et al.*, 1978; Nussinov and Jacobson, 1980; Queen and Korn, 1980). The objectives of most of these programs were straightforward. However, there were some exceptions not included in the survey of Gingeras and Roberts (1980) which were the first attempts to analyze the relationships among different sequences.

The problem of the identification of maximally homologous subsequences among sets of long sequences had been addressed as early as 1966. At that time Fitch (1966), and then Barker and Dayhoff (1972), described heuristic algorithms which attempted to find a means of detecting rigorously the best possible fit between any two matched sequences. The algorithm described by Needleman and Wunsch (1970) was the first to use an iterative matrix method to calculate the relatedness between two sequences. From this approach, Sellers (1974) developed a measure of the drift between sequences. Improvements on the Sellers' algorithm by Waterman *et al*. (1976) allowed for the inclusion of deletions or insertions of variable lengths. Efforts such as these continue in order to calculate the evolutionary distances between sequences. Sellers (1979) has recently provided an alternative approach to this problem by introducing principles from the field of pattern recognition.

The continued development of programs for analyzing nucleotide sequences has grown in the direction of personalization. Modification of existing programs to perform slightly different analytical functions has caused much duplication of effort. This repetition has been fostered partially by the desire to write programs which will run on smaller microcomputers. Thus, many published nucleic acid sequence programs (see Volume 10, Number 1, of Nucleic Acids Research) represent variations of algorithms which have appeared previously. However, there has recently become available several programs which apply a new algorithmic approach to several well-defined problems.

B. Recent Efforts

As the quantity of nucleic acid sequence data grows, two things have become evident. The first is that not all of the information inherent in the genetic code is immediately obvious. The second is that the kinds of questions that can now be addressed have increased due to number of large sequences now collected.

An illustration of these conclusions can be seen in the comparative coding strategies employed by the animal virus genomes of *Simian Virus 40* (*SV40*, 5,243 bp) and *Adenovirus 2* (*Ad2*, 36,500 bp, 70% of which has been sequenced). Although *Ad2* is almost seven times larger than *SV40*, both demonstrate considerable economy in the utilization of sequences for genetic information (Gingeras *et al.*, 1982). This is exemplified by the occurrences of overlapping genes residing at the same position on each of the strands and by the presence of termination codons within the body of polyadenylation signal sequences. While these observations were easily made, the ability to draw the conclusion about efficient use of sequences has depended on the accumulation of a large quantity of sequence information, as well as on the ability to correlate information concerning the map positions, sites and splicing patterns of numerous mRNA species derived from these genomes. At this time, no single computer program could summarize the results of the many types of comparisons and measurements that were performed on these two genomes. Rather, conclusions concerning the similarities of coding strategies were the result of human interpretation of the data supplied by several types of sequence analysis programs. Many of these programs are second generation versions of those listed in Table 1.

1. Sequencing Project Management. Two systems of computer programs have recently emerged from the MOLGEN project at the SUMEX-AIM resources of Stanford University. These systems assist in the organization of DNA sequencing projects. MAXAMIZE is an advisory system which determined, from a predetermined restriction map, the optimal strategy which should be employed in carrying out a sequencing project (Bach *et al.*, 1982). Because of the algorithm employed, projects using the chemical modification (Maxam-Gilbert)

method of sequencing are most suited to this strategy. Another system of programs called GEL takes the data generated during the sequencing project and provides an organizational structure for editing, overlapping and cataloging these data (Clayton and Kedes, 1982). GEL is similar to programs previously described (Gingeras *et al.*, 1979; Staden, 1980b).

Another widely used set of programs for sequence analysis is that originally described by Staden (1977, 1978). He has recently consolidated these programs into a collection which can be used as a comprehensive sequence project management system (Staden, 1980b). This system is aimed at sequencing projects which have accumulated data by means of the so-called "shot-gun" approach. This method of DNA sequencing involves the cleavage of DNA into random fragments, followed by the determination of the sequence of each of these individual fragments. The relationship of the fragments is established by detecting overlaps between pairs of sequences. By gathering sufficient numbers of overlapping segments a complete sequence can be reconstituted. This management system of Staden (1980b) allows for the genealogy of any nucleotide in a collection to be traced back to its individual components. This, in turn, allows the overall quality of the sequence data to be easily assessed.

An alternative management system specifically aimed at sequences collected by the *M13* cloning system (Messing *et al.*, 1981) was recently described by Blumenthal *et al.* (1982). Because *M13* clones contain small inserts of DNA, such clones can be used to provide sequence data and can act as strand-specific single-stranded probes. Consequently, a system of computer programs was designed to keep track of the many clones which would be needed to cover both strands of a large genome. This system consists of three programs linked together to provide a bookkeeping system to detail the location, sequence and origin of each clone. Also described in this collection were several other novel programs to assist in the analysis and display of newly generated sequences.

In addition to these programs designed to assist in the management of sequencing projects, Staden (unpublished) has recently developed a useful companion program called TTRAC which supplements his project management system. TTRAC finds the location of a sequence segment within a longer sequence block by mapping the positions of single nucleotides. Invariably, sequencing experiments produce redundant data, and it is extremely useful to be able to pre-screen templates for redundancy by performing single-channel reactions. By determining the occurrences of any single nucleotide in a sequence, duplicate clones can easily be spotted and eliminated. TTRAC greatly assists in managing a sequencing project by ensuring that the most economical number of sequencing reactions are performed.

2. Sequence Handling and Analysis. The availability of large quantities of sequence data requires utility programs which are capable of sequence data formatting and of performing simple types of analyses. It is this area of sequencing that has the most abundant variety of programs. There are four principal collections of programs widely used for these purposes.

The MRC collection developed by Staden (1977, 1978, 1980a,b), the SEQ collection located in the MOLGEN project at Stanford University (Brutlag *et al.*, 1982) and the collection of Korn and his colleagues (Korn *et al.*, 1977; Queen and Korn, 1980; Sege *et al.*, 1981; Queen *et al.*, 1982) have all undergone continued development from the time of their first release. The results of these efforts are a wide range of programs that will print, annotate, calculate frequencies of occurrences, translate and find symmetries and homologies for sequences of almost any size. The fourth group, the Los Alamos collection (Kanehisa, 1982), has an additional feature of providing a range of subroutines for polypeptide analysis.

A new approach to this type of computer program was described by Schroeder and Blattner (1982). In an effort to develop a collection of programs which would prepare sequence files for analysis by available computer programs, a computer recognizable language (DNA*) has been developed. This language was designed as a shorthand method of manipulating DNA sequences in a vocabulary familiar to molecular biologists. One appealing feature is that the arguments for each operation are written with terms that are used in molecular biology (*e.g.*, AMPGENE, BAM1) so that programs can be written by workers with little or no programming expertise. This type of user-friendly programming language was first developed for molecular biologists by Friedland *et al.* (1982) in a metalanguage called GENGLISH.

The ability to formulate mathematical-like relationships, using the terms and concepts of molecular genetics, represents an important advance. The potential implications of such languages, as used in this and other programs, will be discussed later in this chapter.

3. Prediction Programs. One of the first applications for computers in the analysis of nucleotide sequences was in the prediction of the sequences recognized by restriction endonucleases (Fuchs *et al.*, 1978; Gingeras *et al.*, 1978; Tolstoshev and Blakesley, 1982) and the construction of restriction enzyme maps (Stefik, 1978). For this latter application, an alternative algorithm has recently been formulated in the program MAPCIR-MAPLIN (Pearson, 1982). This program uses data supplied as fragment lengths from single and double restriction enzyme digests to generate maps for linear or circular molecules. Unlike an earlier program, GA1 (Stefik, 1978), which considered all or most of the permutations of a double digest, this program permutes only the possibilities from single digests, resulting in a much faster algorithm.

Another use to which computers have been applied is in identifying protein coding regions in large DNA sequence blocks. The criteria which have been used to make this determination are: (i) the presence of open reading frames (*i.e.*, a string of triplet codons without a termination codon), (ii) the positions of initiation and termination codons, (iii) the presence of potential processing signals such as promoter sites, ribosomal binding sites, splice junctions or polyadenylation sites, and (iv) the presence of codon preference by an organism or cell type. Grantham *et al.* (1980) have proposed that the type of codon preference exhibited by an organism or cell type is common to all the genes expressed in that cell, and thus such preferences can be used to classify cells. Using this and several other assumptions, Staden and McLachlan (1982) have developed a program which predicts the open reading frames that are used by a cell for protein coding. A similar, but more statistical, approach was used by Shulman *et al.* (1981) in their analyses of both RNA (*MS2*) and DNA (*ϕX174*) genomes.

An alternative to the codon preference approach was described by Shepherd (1981), who has studied the pattern of purine and prymidine occurrences in both eukaryotic and prokaryotic genomes. His observations are consistent with the hypothesis that primitive messages were formed of coding triplets having the form PNY (P = purine, N = any nucleotide, Y = pyrimidine). As a result, identification of reading frames used for protein coding closely follows the codon choice that most resembles this primeval sequence.

Computer programs have also been used to analyze sequence data so as to generate biologically meaningful predictions in more complex situations, using algorithms borrowed from the field of pattern recognition. Based on the pioneering works of Rosenblatt (1950) and Minsky and Papert (1969), a class of feature detection systems having self-teaching capabilities has been developed. These systems were termed "perceptrons". Using a perceptron algorithm, Stormo, Schneider and Gold (1982) have developed computer programs which attempt to predict translational initiation sites (ribosomal binding sites).

The first rules for the localization of a ribosomal binding site in a nucleotide sequence were defined by Shine and Dalgarno (1974). However, it quickly became clear that not all legitimate mRNA sequences contained regions which obeyed these rules. By analyzing a large number of defined prokaryotic mRNA sequences, the number and types of rules were increased so as to improve the likelihood that legitimate coding sequences without Shine-Dalgarno regions could be detected by computer analysis (Stormo, Schneider and Gold, 1982). By borrowing principles from the perceptron system, an alternative algorithm to this "rule approach" was developed. The algorithm scans linear sequences and defines a probability function (a weighting function) which will give a value to a particular sequence as a potential ribosomal binding site (Stormo, Schneider, Gold and Ehrenfeucht, 1982). There are two advantages of this approach over a consensus approach. Firstly, little or nothing is specified about the sequences being tested, except that they are part of some functional set. Thus, no single feature is absolutely required. Secondly, each site in a sequence is quantitatively evaluated by virtue of its weighted score. Therefore, there can be an ordering of potential sites based on their weighted values which may eventually reflect the inherent "strengths" as ribosomal binding sites.

Significantly, this perceptron approach and programs used in the MOLGEN-GENESIS system (Friedland *et al.*, 1982) are the first applications of elements from the field of artificial intelligence into computer programs for the analysis of nucleic acid sequences. The use of self-learning algorithms from the field of artificial intelligence is important because there is accumulating evidence that simple rules are not applicable in defining important biological signals, such as ribosomal binding sites, promoter sites and splice sites. Rather, a learning process will be required before it will be possible to predict whether potential sites are more or less likely candidates. Learning algorithms such as perceptrons have already suggested that selected nucleotide sequences are not the only requirement for certain biological signals.

4. Other Areas of Interest. Table 2 summarizes some of the important elements of programs developed over the last two years. In addition to the programs mentioned above, several others have been developed to study topics such as secondary structure formation, sequence homology and production of knowledge organization systems. Because the accumulation of isolated facts can often be as confusing as it can be helpful, the development of knowledge organization systems is likely to become increasingly important.

III. ORGANIZATION OF INFORMATION

A. Data Base Management Systems

The establishment of computerized systems which are able to keep track of and correlate detailed pieces of data has become necessary in a wide variety of fields. To accomplish this task, automated systems have been developed and are called data base management systems (DBMSs). The essential elements of such systems are a variety of information files (a data base) which can be linked together to form information files to highlight user-created relationships among the elements present in the data files. DBMS refers to the set of programs that accomplish the tasks of organization and maintenance of these files.

The presence of both sequence and nonsequence data within the structure of a DBMS offers several attractive features, including: (i) *Multiple views of data* DBMSs provide a means of easily reformatting and recombining data extracted from central pools. (ii) *Data Base Flexibility*. Frequently, once a particular program has been written, the results lead to statements such as, "It would be nice if this program could do . . .". DBMSs frequently allow users to create new utilities. (iii) *Nonredundancy of Data*. Often, one piece of information appears in several different data files (*e.g.*, restriction enzyme sites). If sequences are altered, then every occurrence of that piece of information (*e.g.*, positions) must be updated. Thus, because any piece of data usually appears

Table 2. Recently developed programs for DNA sequence analysis.

TITLE	FUNCTION	SOURCE	LANGUAGE
I. Sequencing Project Management			
MAXAMIZE	From RE map, sequencing strategy suggested for Maxam-Gilbert method.	Friedland *et al.* (1982) (MOLGEN, SUMEX: Stanford U.)	Interlisp (GENGLISH)
GEL	System for entering, editing and generation of consensus overlaps for DNA sequence.	Clayton and Kedes (1982) (MOLGEN, SUMEX: Stanford U.)	SAIL
TTRAC	Denotes overlaps between two sequences using single nucleotide track.	Staden (unpublished) (MRC, Cambridge)	FORTRAN
noname	System to organize data shotgun strategy. Genealogy of each nucleotide recorded.	Staden (1980b) (MRC, Cambridge)	FORTRAN
M13,SEQ, FIND	Bookkeeping programs for *M13* cloning sequencing system.	Blumenthal *et al.* (1982) (Cold Spring Harbor)	FORTRAN
II. Sequence Handling and Analysis			
SEQ	Interactive programs to manipulate, format and perform primary analysis of sequence data.	Brutlag *et al.* (1982) (MOLGEN, SUMEX: STANFORD U.)	SAIL
Los Alamos Collection	Similar to SEQ, with polypeptide analysis subroutines.	Kanahisa (1982) (Los Alamos)	FORTRAN
noname	Improved system of programs to manipulate and analyze nucleic and amino acids.	Queen *et al.* (1982) (NIH)	PL/I
DNA*	META-language for editing and manipulating sequences.	Schroeder and Blattner (1982) (U. of Wisconsin)	DNA*
III. Best Fit Programs			
Homology Search	Improved program to find optimum homologies between nucleotide sequence sets.	Queen *et al.* (1980) (NIH)	PL/I

Table 2 (continued)

TITLE	FUNCTION	SOURCE	LANGUAGE
noname	Uses Needleman-Wunsch-Sellers algorithm to find all subsequences in a long sequence which resemble each other.	Goad and Kanehisa (1982) (Los Alamos)	FORTRAN
IV. Secondary Structure			
noname	Programs in this section provide a collection of	Kanehisa and Goad (1982) (Los Alamos)	FORTRAN
noname	stable structures, which can be assessed with RNA, based	Dumas and Ninio (1982) (IRBM, Paris)	FORTRAN
noname	on ability to find optimum number and sizes	Studnicka *et al.* (1982) (UCLA)	APL
noname	of homologous subsequences within a sequence.	Zuker and Stiegler (1982) (PROPHET, Cambridge, Mass., and NRC, Ottawa)	FORTRAN
V. Prediction Programs			
Perceptron	Predicts 5'-ends of genes.	Stormo *et al.* (1982) (U. of Colorado)	PASCAL
MAPCIR-MAPLIN	Constructs restriction site maps.	Pearson (1982) (Johns Hopkins U.)	FORTRAN
noname	Predicts protein coding regions after identifying coding preference of genome.	Staden and McLaughlan (1982) (MRC, Cambridge)	FORTRAN
GENESIS	These last three programs store information within a hierarchically organized set of	Friedland *et al.* (1982) (MOLGEN, SUMMEX: Stanford U.)	Interlisp (GENGLISH)
LIBRARIAN	of structures. The programs are used to relate stored data	Schneider *et al.* (1982) (U. of Colorado)	PASCAL (DELILA)
National Biomedical Research Collection	elements.	Orcutt *et al.* (1982) (National Biomedical Research Laboratory)	---

only once in a DBMS, updating becomes easier and inconsistencies in the data base are avoided. (iv) *Self-expansion of Data Bases.* The results of new constructs and searches can be included automatically as elements of the data base (see example below).

Perhaps, at this point, an example would aid in understanding. The sequence shown in Figure 2 consists of 4,000 nucleotides (~11%) from the left end of the *Ad2* genome (Gingeras *et al.*, 1982). One way to organize this information is to ask which areas of this sequence are potential coding regions. There are two major divisions possible in this functional approach: those regions of the sequence which, from independent data, are known to encode genetic information, and those regions of the sequence which potentially can be used for coding but for the time being remain as unidentified open reading frames (URFs). Features like the positions of the 5' end cap sites, the first AUG codon used as an initiator, splice sites, termination codons, polyadenylation signal sites and the 3' end of each real or potential coding region are routinely recorded. As can be seen from Figure 2b, there are six open reading frames capable of encoding proteins of greater than 10,000 daltons. In addition to the features mentioned above, observations from such areas as electron microscopy, mutation studies and protein chemistry could also be recorded to help in determining if any of the URFs are actually used.

Besides this segment of the genome, there are another 32,000 nucleotides of sequence which comprise the rest of the *Ad2* genome. Within these sequences there are several dozen URFs of sizes equal to, or larger than, the two observed at the left end. Each of these reading frames can be analyzed for the same features as were noted for the first six reading frames. Added to this is the fact that there are many hundreds of thousands of sequenced nucleotides determined from other genomes, so that the amount of potential coding information that can be cataloged is considerable. In what way can a DBMS bring new insight into the analysis of this data?

Consider the phenomenon of RNA splicing (Berget *et al.*, 1977; Chow *et al.*, 1977). By this process, noncontiguous pieces of RNA transcribed from a genome can be linked together to form a message in chain-like structure. Recently, it was reported that, in several cases, the segments of RNA (introns) removed from the final transcript can be found as closed circular structures (Arnberg *et al.*, 1980; Grabowski *et al.*, 1981). There has also been evidence that usable genetic information is sometimes present within introns (Lazowska *et al.*, 1980; Gingeras *et al.*, 1982). One possibility is that there may be novel coding strategies built into this system of RNA processing. Figure 3 illustrates that, after a circular piece of RNA has been constructed from the intron of a spliced mRNA, there is no longer a true 5' or 3' end to this molecule. Consequently, a single strand scission made at any other position, aside from the initial point of ligation, will result in a construct with a new 5' terminus which may be capable of being translated into a variety of protein products, one of which (at the junction point) is entirely novel. Of course, peptides from such RNA constructs will have a characteristic set of sizes and amino acid compositions. These features, as we have seen above, can be cataloged by simple computer searches. Thus, for any particular spliced mRNA, each intron region can be scanned for all possible open reading frames in order to determine the size and amino acid sequence for each putative peptide. These computer generated candidates can be compared to the peptides that have been observed from *in vivo* results to determine if such novel coding strategies are indeed employed.

B. Available Systems

There are three principal DBMSs currently utilized for nucleic acid sequence storage and analysis. The oldest of these is run by the National Biomedical Research Foundation. Until recently, their concern had been with the collection and correlation of protein sequence information. In 1980, however, a nucleic acid data base was instituted (Orcutt *et al.*, 1982). This DBMS is directed toward providing complete bibliographic background for all collected

2a

```
ADENOVIRUS 0-11% (JAN 13 ,82)
CATCATCATAATATACCTTATTTTGGATTGAAGCCAATATGATAATGAGG
GGGTGGAGTTTGTGACGTGGCGCGGGGCGTGGGAACGGGGCGGGTGACGT
AGTAGTGTGGCGGAAGTGTGATGTTGCAAGTGTGGCGGAACACATGTAAG
CGCCGGATGTGGTAAAAGTGACGTTTTTGGTGTGCGCCGGTGTATACGGG
AAGTGACAATTTTCGCGCGGTTTTAGGCGGATGTTGTAGTAAATTTGGGC
GTAACCAAGTAATGTTTGGCCATTTTCGCGGGAAAACTGAATAAGAGGAA
GTGAAATCTGAATAATTCTGTGTTACTCATAGCGCGTAATATTTGTCTAG
GGCCGCGGGGACTTTGACCGTTTACGTGGAGACTCGCCCAGGTGTTTTTC
TCAGGTGTTTTCCGCGTTCCGGGTCAAAGTTGGCGTTTTATTATTATAGT
CAGCTGACGCGCAGTGTATTTATACCCGGTGAGTTCCTCAAGAGGCCACT
CTTGAGTGCCAGCGAGTAGAGTTTTCTCCTCCGAGCCGCTCCGACACCGG
GACTGAAAATGAGACATATTATCTGCCACGGAGGTGTTATTACCGAAGAA
ATGGCCGCCAGTCTTTTGGACCAGCTGATCGAAGAGGTACTGGCTGATAA
TCTTCCACCTCCTAGCCATTTTGAACCACCTACCCTTCACGAACTGTATG
ATTTAGACGTGACGGCCCCCGAAGATCCCAACGAGGAGGCGGTTTCGCAG
ATTTTTCCCGAGTCTGTAATGTTGGCGGTGCAGGAAGGGATTGACTTATT
CACTTTTCCGCCGGCGCCCGGTTCTCCGGAGCCGCCTCACCTTTCCCGGC
AGCCCGAGCAGCCGGAGCAGAGAGCCTTGGGTCCGGTTTCTATGCCAAAC
CTTGTGCCGGAGGTGATCGATCTTACCTGCCACGAGGCTGGCTTTCCACC
CAGTGACGACGAGGATGAAGAGGGTGAGGAGTTTGTGTTAGATTATGTGG
AGCACCCCGGGCACGGTTGCAGGTCTTGTCATTATCACCGGAGGAATACG
GGGGACCCAGATATTATGTGTTCGCTTTGCTATATGAGGACCTGTGGCAT
GTTTGTCTACAGTAAGTGAAAATTATGGGCAGTCGGTGATAGAGTGGTGG
GTTTGGTGTGGTAATTTTTTTTTTAATTTTTACAGTTTTGTGGTTTAAAGA
ATTTTGTATTGTGATTTTTTTAAAAGGTCCTGTGTCTGAACCTGAGCCTGA
GCCCGAGCCAGAACCGGAGCCTGCAAGACCTACCCGGCGTCCTAAATTGG
TGCCTGCTATCCTGAGACGCCCGACATCACCTGTGTCTAGAGAATGCAAT
AGTAGTACGGATAGCTGTGACTCCGGTCCTTCTAACACACCTCCTGAGAT
ACACCCGGTGGTCCCGCTGTGCCCCATTAAACCAGTTGCCGTGAGAGTTG
GTGGGCGTCGCCAGGCTGTGGAATGTATCGAGGACTTGCTTAACGAGTCT
GGGCAACCTTTGGACTTGAGCTGTAAACGCCCCAGGCCATAAGGTGTAAA
CCTGTGATTGCGTGTGTGGTTAACGCCTTTGTTTGCTGAATGAGTTGATG
TAAGTTTAATAAAGGGTGAGATAATGTTTAACTTGCATGGCGTGTTAAAT
GGGGCGGGGCTTAAAGGGTATATAATGCGCCGTGGGCTAATCTTGGTTAC
ATCTGACCTCATGGAGGCTTGGGAGTGTTTGGAAGATTTTTCTGCTGTGC
GTAACTTGCTGGAACAGAGCTCTAACAGTACCTCTTGGTTTTGGAGGTTT
CTGTGGGGCTCCTCCCAGGCAAAGTTAGTCTGCAGAATTAAGGAGGATTA
CAAGTGGGAATTTGAAGAGCTTTTGAAATCCTGTGGTGAGCTGTTTGATT
CTTTGAATCTGGGTCACCAGGCGCTTTTCCAAGAGAAGGTCATCAAGACT
TTGGATTTTTCCACACCGGGGCGCGCTGCGGCTGCTGTTGCTTTTTTGAG
TTTTATAAAGGATAAATGGAGCGAAGAAACCCATCTGAGCGGGGGGTACC
TGCTGGATTTTCTGGCCATGCATCTGTGGAGAGCGGTGGTGAGACACAAG
AATCGCCTGCTACTGTTGTCTTCCGTCCGCCCGGCAATAATACCGACGGA
GGAGCAACAGCAGGAGGAAGCCAGGCGGCGGCGGCGGCAGGAGCAGAGCC
CATGGAACCCGAGAGCCGGCCTGGACCCTCGGGAATGAATGTTGTACAGG
TGGCTGAACTGTTTCCAGAACTGAGACGCATTTTAACCATTAACGAGGAT
GGGCAGGGGCTAAAGGGGGTAAAGAGGGAGCGGGGGGCTTCTGAGGCTAC
AGAGGAGGCTAGGAATCTAACTTTTAGCTTAATGACCAGACACCGTCCTG
AGTGTGTTACTTTTCAGCAGATTAAGGATAATTGCGCTAATGAGCTTGAT
CTGCTGGCGCAGAAGTATTCCATAGAGCAGCTGACCACTTACTGGCTGCA
GCCAGGGGATGATTTTGAGGAGGCTATTAGGGTATATGCAAAGGTGGCAC
TTAGGCCAGATTGCAAGTACAAGATTAGCAAACTTGTAAATATCAGGAAT
TGTTGCTACATTTCTGGGAACGGGGCCGAGGTGGAGATAGATACGGAGGA
TAGGGTGGCCTTTAGATGTAGCATGATAAATATGTGGCCGGGGGTGCTTG
GCATGGACGGGGTGGTTATTATGAATGTGAGGTTTACTGGTCCCAATTTT
AGCGGTACGGTTTTCCTGGCCAATACCAATCTTATCCTACACGGTGTAAG
CTTCTATGGGTTTAACAATACCTGTGTGGAAGCCTGGACCGATGTAAGGG
TTCGGGGCTGTGCCTTTTACTGCTGCTGGAAGGGGGTGGTGTGTCGCCCC
AAAAGCAGGGCTTCAATTAAGAAATGCCTGTTTGAAAGGTGTACCTTGGG
TATCCTGTCTGAGGGTAACTCCAGGGTGCGCCACAATGTGGCCTCCGACT
GTGGTTGCTTCATGCTAGTGAAAAGCGTGGCTGTGATTAAGCATAACATG
GTGTGTGGCAACTGCGAGGACAGGGCCTCTCAGATGCTGACCTGCTCGGA
CGGCAACTGTCACTTGCTGAAGACCATTCACGTAGCCAGCCACTCTCGCA
AGGCCTGGCCAGTGTTTGAGCACAACATACTGACCCGCTGTTCCTTGCAT
TTGGGTAACAGGAGGGGGGTGTTCCTACCTTACCAATGCAATTTGAGTCA
CACTAAGATATTGCTTGAGCCCGAGAGCATGTCCAAGGTGAACCTGAACG
GGGTGTTTGACATGACCATGAAGATCTGGAAGGTGCTGAGGTACGATGAG
ACCCGCACCAGGTGCAGACCCTGCGAGTGTGGCGGTAAACATATTAGGAA
CCAGCCTGTGATGCTGGATGTGACCGAGGAGCTGAGGCCCGATCACTTGG
TGCTGGCCTGCACCCGCGCTGAGTTTGGCTCTAGCGATGAAGATACAGAT
TGAGGTACTGAAATGTGTGGGCGTGGCTTAAGGGTGGGAAAGAATATATA
AGGTGGGGGTCTCATGTAGTTTTGTATCTGTTTTGCAGCAGCCGCCGCCA
TGAGCGCCAACTCGTTTGATGGAAGCATTGTGAGCTCATATTTGACAACG
CGCATGCCCCCATGGGCCGGGGTGCGTCAGAATGTGATGGGCTCCAGCAT
TGATGGTCGCCCCGTCCTGCCCGCAAACTCTACTACCTTGACCTACGAGA
CCGTGTCTGGAACGCCGTTGGAGACTGCAGCCTCCGCCGCCGCTTCAGCC
GCTGCAGCCACCGCCCGCGGGATTGTGACTGACTTTGCTTTCCTGAGCCC
GCTTGCAAGCAGTGCAGCTTCCCGTTCATCCGCCCGCGATGACAAGTTGA
CGGCTCTTTTGGCACAATTGGATTCTTTGACCCGGGAACTTAATGTCGTT
TCTCAGCAGCTGTTGGATCTGCGCCAGCAGGTTTCTGCCCTGAAGGCTTC
CTCCCCTCCCAATGCGGTTTAAAACATAAATAAAAACCAGACTCTGTTTG
```

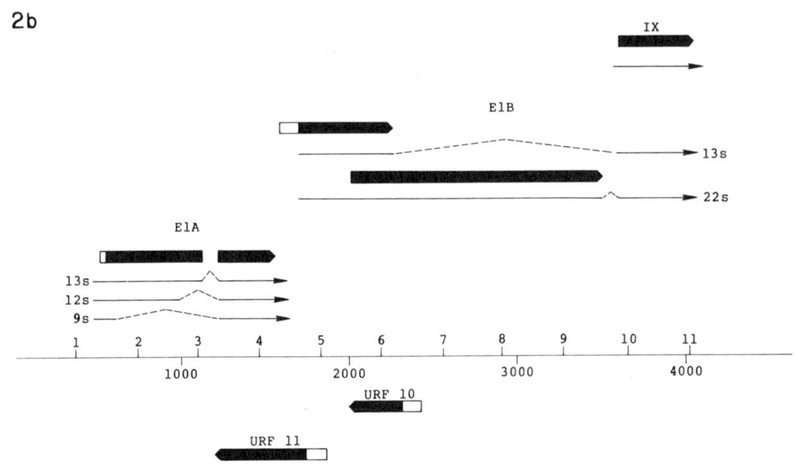

Figure 2. (a) The first 4050 nucleotides from the left end of the *Ad2* virus genome. Contained within this collection of nucleotides are the instructions for the synthesis of seven previously observed proteins and the potential for at least another two (Gingeras *et al.*, 1982). (b) The available coding regions found within the sequence presented in Figure 2a. Such coding regions are called open reading frames. Regions E1A, E1B and IX have been shown to transcribe mRNA species. Filled-in sections of arrows start with an initiation codon (AUG), and open sections indicate preceding open reading frames which contain no proceeding initiation codons. Splicing patterns are represented by broken lines. The L-strand contains two open reading frames which as yet have not been shown to be transcribed (URF: unidentified reading frame).

sequences, as well as supplying some analysis capability for user-supplied or previously stored sequences. In its analysis efforts, this DBMS resembles the sequence handling and management systems discussed in an earlier section of this review (Queen and Korn, 1980; Brutlag *et al.*, 1982; Kanehisa, 1982). One advantage that this system possesses is the availability of over 550,000 nucleotides of stored sequence, which probably represents the largest single collection of nucleic acid sequence data in computer readable form.

Two alternative approaches to the concept of DBMSs for nucleic acid sequences have recently been described. The first of these is a system called LIBRARIAN (Schneider *et al.*, 1982). Like the National Biomedical Research Foundation system, LIBRARIAN contains

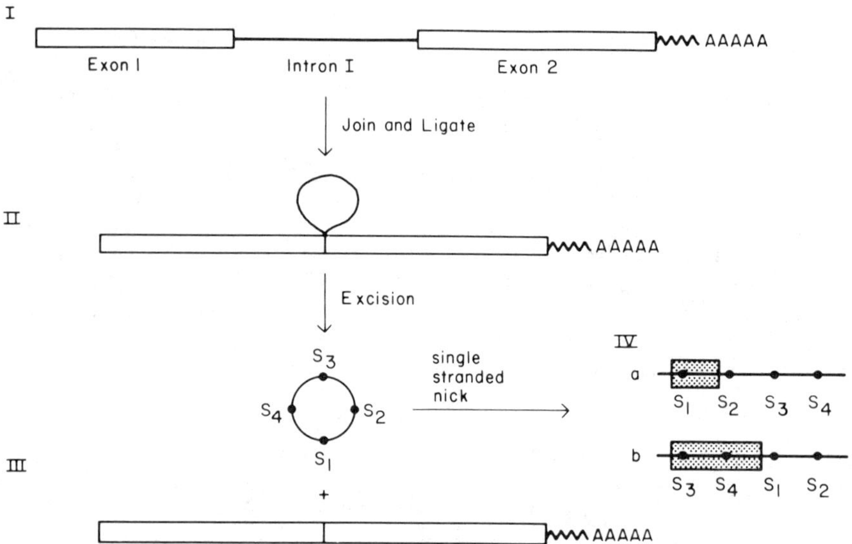

Figure 3. A hypothetical representation of the possible increase in coding capacity achieved by means of a circular intron. After the intron is removed as a closed circle (III), there are four possible (S1, S2, S3, S4) start codons (AUG) which could be used for translation. A single-stranded nick in the circle between any of the four start codons would render one the 5' AUG of a new mRNA. Only two of the possible products are shown in stage IV. IVa is an mRNA which has S1 as the start codon with an open reading frame extending almost to S2. IVb is an mRNA that has been formed by a "nick" between S2 and S3. This allows S3 to operate as the initiation codon with an open reading frame extending almost to S1.

a large collection of stored sequence information as well as automated utilities to search and analyze these data. However, this DBMS is aimed at the need to manage locally controlled libraries. To accomplish this, LIBRARIAN uses a language called DELILA, devised to assist in manipulating sequences from the library of data by English-like commands. Because the information stored in the library contains biological relationships as well as sequences, this language allows the labelling (and thus retrieval) of specific segments of the sequence library by using biological functions as labels. In so doing, it can act in larger computers like a standard editor in much the same way that the DNA* language (Schroeder and Blattner, 1982) operates.

The use of such languages as DELILA and DNA* is the first step in providing a means of establishing mathematical-like relationships for sequence analysis. Such relationships are very important since most biological information is essentially descriptive. Because sequence data is quantitative, these languages provide a means of transferring phenomenological data into the realm of quantitative information. By using such languages, the formulation of relationships can be worked out in a machine-understandable form containing concepts familiar to molecular biologists. One consequence is that it is now possible for computer-naive scientists to write a wide variety of personalized computer programs. An example of such a program is listed in Figure 4.

Given the continual advancement of technology within the computer industry, one could foresee the production of small but powerful personal computers specifically designed for molecular biologists, composed of logic boards with the essential elements of appropriate languages hard-wired into them. These processors

```
COMMENT:  Select from the Ad2 virus sequence the coding portions
          (minus the introns) of two mRNAs from the transforming
          region of the genome, and clone the smallest message
          into the BamHI site of pBR322 by G,C tailing.

1.  MRNA1 = AD2.SEQ (559>973)(1226>1540) + 90* "A"

2.  MRNA2 = AD2.SEQ (559>1111)(1226>1540) + 90* "A"

3.  REVBAMPBR = PBR322.SEQ (375<376)

4.  CLONE = REVBAMPBR + 10* "G" + MRNA1 + 10*  ~"G"
```

Figure 4. A computer program written in a molecular biology oriented computer language called DNA* (Schroeder and Blattner, 1982). The result is a computer generated circular hybrid sequence which can be stored as part of the library of sequence data files such as PBR322.SEQ and Ad2.SEQ1. Symbols used in this program have the same meaning as those listed by Schroeder and Blattner (1982).

would provide analysis functions for newly entered or pre-stored sequence information in the way that pocket calculators do for users interested in both general and specialized mathematical operations in fields such as accounting, engineering and actuarial science.

The second interesting approach to the structure of DBMSs has been suggested by Friedland *et al.* (1982). These authors make the important distinction between a DBMS and a Knowledge Base Management System (KBMS). Included in their discussion of the distinctions between data bases and knowledge bases are such factors as complexity of information, symbolic versus numeric forms of information, variety of information and the presence of higher-order knowledge. Central to their discussion are their comments on the nature of the information of interest to, and stored by, molecular biologists. Because the types of information stored in molecular biological data bases are frequently in the form of concepts (promoters, ribosomal binding sites, origins of replication and splice points), the use of these collections is frequently aimed at developing strategies designed to formalize this data. Taking into account the sometimes ill-defined nature and the intended purposes for which this data is to be used, a knowledge-based genetic engineering simulation system called GENESIS has been developed to represent both factual and conceptual information. By storing information within a hierarchically organized framework, user-created compact pools of knowledge can be created by specialized searches. This framework is composed of information cells called *units* which provide a cataloging mechanism for the management of an ever increasing data base. Data collected in these cells can be transformed into "knowledge" by use of any of three symbolic representation systems called MAPS, SEQUENCES and RULES. The first two systems act by means of sophisticated editors to allow a user to represent sequence and biological data in the form of annotated sequence displays, linear or circular maps. Any of the data used in these representations can be supplemented with complete bibliographical and genetic references.

The RULES representational system is a tool to allow molecular biologists to define rules governing the relationships that connect the data stored in any of the *units*. Operations such as join (LIGATE), excise (DELETE), insert (ADD), transcribe and splice can be applied to any sequence in storage. The goal of defining such relationships is to permit researchers to simulate experiments, design strategies and catalog predictable results. To accomplish goals, operations defined by a user are expressed in another English-like meta-language called GENGLISH. In this respect, the GENESIS system is similar to the DELILA and DNA* systems. The differences reside in the fact that GENGLISH is not a machine readable language (as are DELILA and DNA*). Rather, the GENESIS system works by using a formal computer language called INTERLISP. Thus, "programming" in GENGLISH can be done only within the context of the GENESIS system.

Recently, the National Institutes of Health (NIH) in the United States and the European Molecular Biology Laboratory (EMBL) in Heidelberg, Germany, have taken the first steps in creating centralized nucleotide sequence data bases. The NIH award was made to Bolt, Beranek and Newman of Cambridge, Massachusetts, which is a company with expertise in computer communications. With help from the Los Alamos National Laboratory, they will be responsible for compiling and distributing a national data base of nucleotide sequences. At the EMBL, the Nucleotide Sequence Data Laboratory was created in early 1982 to provide a similar service for European laboratories. As valuable as these data bases will be, they have yet to be funded for the development of sophisticated sequence manipulation programs. Such program development would seem to be the most likely next step, and the establishment of these facilities clearly points out the need to organize the rapidly growing body of sequence data.

The combination of a well organized data base and the means to manipulate this information creatively for sequence analysis within a KBMS may be one of the most important contributions computers have made to molecular biology. Because naive computer users can write their own analysis utilities by using GENESIS, LIBRARIAN or DNA*, the promise of new insights into the more subtle regulatory processes of molecular biology is almost certain.

IV. STRUCTURE AND FUNCTION: THE NEXT GENERATION OF COMPUTER PROGRAMS

A. Description of the Problem

Studies concerning the way nucleic acid molecules bend and fold into secondary and tertiary structures have long been considered a rather esoteric subject. This remains true even though the three-dimensional structure of nucleic acid molecules has been implicated as being important in such biological processes as DNA replication, transcription initiation and termination, RNA splicing, ribosomal assembly and protein translation. The relative absence of well defined rules governing the formation of such structures is one roadblock to further progress in this area. The first attempts to establish a descriptive set of rules concerned the self-folding secondary structures that nucleic acid molecules can assume (Tinoco *et al.*, 1971). This, and subsequent approaches, used information such as base stacking, hydrogen bonding and loop free energy data to measure the physico-chemical properties of oligomers and homopolymers (Salser, 1977).

Computer-assisted prediction of the secondary structures of RNA molecules was first attempted by Pipas and McMahon (1975). Since then, similar programs have been devised to construct the best free energy models based on primary sequence data (Studnicka *et al.*, 1978; Zuker and Stiegler, 1981). The algorithms used represent slightly different approaches which have recently been reviewed (Auron *et al.*, 1982). The validity of computer generated models, produced by any of these programs, can be tested with chemical or enzymatic probes.

Such enzymatic probes include ribonucleases *A*, *T1*, *T2*, *S1* and mung bean nuclease which are used to cleave accessible single-stranded regions of RNA. Double-stranded regions of RNA can be detected by use of the RNAse activity from *Naja naja oxiana* (cobra) venom (Vassilenko and Babkina, 1965). Chemical modification probes such as guanosine-specific kethoxal (Staehelin, 1959; Shapiro and Hachmann, 1966) or chemical cross-linking agents such as psoralen (Wollenzien *et al.*, 1979) can also be used to evaluate the enzymatic data and obtain limited information concerning tertiary structure.

With the exception of certain tRNA molecules, the use of X-ray crystallographic methods for structural studies of larger nucleic acid molecules has been impossible due to the scarcity of large enough quantities of highly purified samples. Information on the solution structure of nucleic acid molecules is thus often derived from the combination of computer generated models and nuclease studies.

Such an approach has been described in two sets of studies on both prokayotic (Douthwaite and Garrett, 1981; Douthwaite *et al.*, 1982) and eukaryotic (Pavlakis *et al.*, 1979, 1980) RNA structures. In these studies RNA molecules are labelled with P-32 at either the 5' or the 3' termini, followed by partial cleavage or chemical modification with the probes described previously. In the Douthwaite studies, only single site cleavage products were isolated prior to denaturation and resolution on urea-containing polyacrylamide gels. The result of these reactions is the generation of large amounts of detailed information concerning the susceptible sites within the RNA structure. This information is in the form of bands on a polyacrylamide gel, resembling end-labelled fragments from a partial restriction endonuclease digestion. Thus, an autoradiograph may contain a hundred or more bands, which must be interpreted to extract data concerning the location of single- or double-stranded regions, the location of secondary sites of cleavage or modification which become available only after the effects of a primary cleavage, and the location of appropriate substrate sites which are hidden from the activity of probes. It is with this reservoir of data that computers may be of most assistance.

B. Possible Approach

During the process of correlating nuclease data with features from computer generated models, the location of single- and double-stranded regions, as well as the results of a partial sequence map are compared. The derivation of the computer model is not linked to experimental data. Usually, only during editing and subsequent pruning of the models do the effects of the experimental data become important. An alternative system could be conceived which relies on the information obtained from nuclease and modification experiments before the formation of any model.

Figure 5 contains a model of the 159 nucleotide *VAI* RNA from *Ad2* virus (Zain *et al.*, 1979). It is the computer representation of the structure of this molecule based on the nucleotide pairing rules described by Pipas and McMahon (1975). Another approach to a more realistic model could include the characteristics of various nuclease treatments of the *VAI* structure. For any nucleotide, a determination of + or - would exist for its susceptibility to any of the probes used. This table of results could then be used as a list of rules (example: A36, U101, A103, C104, G106 must be left in single-stranded form) to construct a set of models. These models would be formed by using data from both nuclease and thermodynamic approximations.

There are two obvious advantages to this approach. Firstly, the organization of experimental data concerning the properties of solution structures can be ensured, thus providing for the fidelity of this information. This makes it potentially profitable for additional analyses to be performed leading to the discovery of the principles which govern the formation of higher-order structure. Secondly, these models can provide information about the properties of the probes used in such studies. Since it is apparent that both nucleases and chemical modifying agents have preferences as to which sites are cleaved or labelled, it is important to understand what those preferences are. Records concerning the activities of each of these probes on various substrates will help to achieve this goal. This last point is important because in some instances

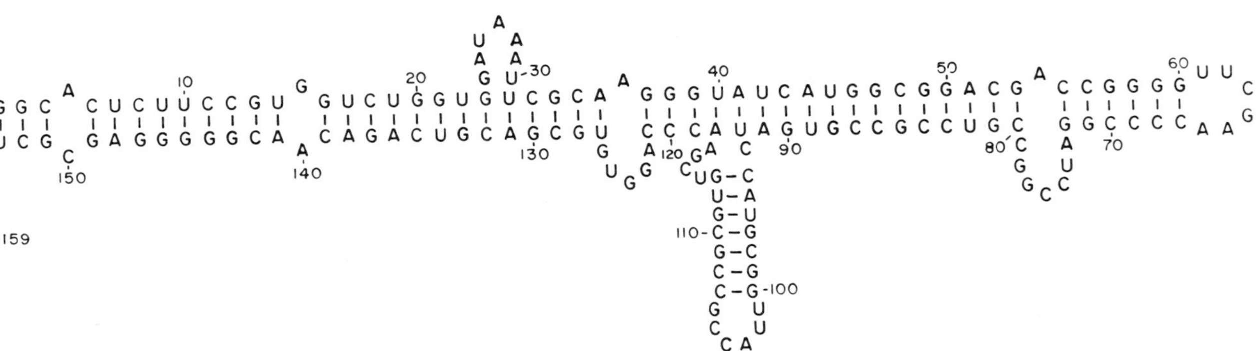

Figure 5. A secondary structure model of *VAI* RNA (Zain *et al*., 1979) which was produced by using the base pairing and folding rules of Pipas and McMahon (1975). The total minimum free energy calculation is expressed as kcal/mole.

it is difficult to decide whether the results observed for secondary structure studies reflect information about the properties of the substrate or about the probe.

The subtle rules which govern the way form dictates activity remain a highly challenging problem in molecular biology. One of the first steps towards understanding this phenomenon is the possession of a means of representing molecular structures accurately. Already it is clear that the form and stability of higher structures is influenced by many factors. Computer-assisted methods provide tools and organization for those interested in these studies.

V. CONCLUSION

While computers are now accepted as the most efficient means of organizing and deciphering information from the growing body of DNA sequence data, a more complex role for computers has recently started to emerge. It is becoming clear that certain biologically important control signals cannot be identified by using the earliest computer programs. By borrowing principles from the field of artificial intelligence, an effort is being made to take advantage of the large body of accumulated sequence data to find patterns which specify the control of genetic expression. In much the same way as the human mind seeks out patterns from an abstract painting, computers are being taught to seek out regulatory "images" from large collections of sequence data. However, as in art, such interpretations are subject to criticism. Therefore, each computerized review of a sequence will ultimately have to await the test of experimentation.

Other efforts, such as the development of computer languages using terms and constructs common to molecular biologists, signal the continuing evolution of the union of computers and molecular biology. Future use of the computer will undoubtedly include the reduction of large collections of detailed biological or sequence information into concise graphic forms. If the understanding of biological processes indeed depends on the correlation of multiple and varied pieces of data, then use of computer graphics will undoubtedly be important in the representation of stored or processed data.

ACKNOWLEDGMENTS

I wish to thank Dr. R. J. Roberts and Dr. J. E. Brooks for their helpful discussions and critical reading of this manuscript. I am grateful to N. D'Anna, D. Blake and E. Lockhart for their help in preparing this manuscript. This work was supported by Grant CA27275 from the National Institutes of Health to T. R. Gingeras and R. J. Roberts.

Chapter 3

THE ROLE OF MODELS IN THE ANALYSIS OF MOLECULAR GENETIC DATA, WITH PARTICULAR REFERENCE TO RESTRICTION FRAGMENT DATA

W. J. Ewens

Department of Biology, University of Pennsylvania, Philadelphia, Pennsylvania

and

Department of Mathematics, Monash University, Clayton, Victoria, Australia

I. INTRODUCTION

The restriction endonuclease technique introduces several problems of a theoretical nature. Here we take up two of these. The first problem concerns the estimation of some genetic quantity (*e.g.*, degree of heterozygosity). Any estimation procedure is an inference requiring some assumptions to be made on the generation of the data. Which are the most reasonable assumptions, or the minimal set of assumptions, is not always clear, nor is it clear whether or not genetic models are needed. Here we introduce various sets of assumptions and investigate the estimators to which they lead.

The second problem concerns standard errors of estimates. A standard error measures the degree of variation in the estimate in conceptual repetitions of the experiment. However, various concepts of repetition are possible, and for different purposes different concepts are preferable. Quite large differences in standard deviation estimates are possible using different points of view. The two concepts of repetition we consider are (i) repetition of the entire evolutionary process (or, equivalently, repetition by using a different restriction enzyme, or repetition by moving to a different part of the genome), and (ii) repetition by sampling different individuals from the same population. The standard errors are much smaller under the latter concept than under the former. Which concept is to be preferred will depend on the aim of the investigation.

II. MEASURING GENETIC VARIATION

Perhaps the two questions most commonly asked in empirical and theoretical population genetics are, respectively, "How much genetic variation is there in natural populations?" and "What is the cause of this variation?". The first question appears to be entirely empirical, but is not really so: we will seldom be in a position to survey an entire population, and, if we can, will usually survey only a fraction of the genome. To make inferences from the sample to the entire population requires at least some theoretical development. This might be at the comparatively simple level of standard sampling theory, but in some cases, where the method of deriving the data is not a standard one, more complex theory might be required.

This chapter considers such complex theory arising when restriction endonucleases are used to provide partial evidence on the degree of genetic variation in a sample from a population. We wish to use this partial information to estimate commonly employed measures (*e.g.*, degree of heterozygosity) of variation. We will assume throughout the genetic treatment that the variation is caused by mutation and random genetic drift in a selectively neutral set of sites in the genome.

There are two concepts of genetic variation of both practical and theoretical interest. The first is a simple one: "How much genetic variation is there present in this generation of this population at the part of the genome surveyed?" Here other generations of the population and other parts of the genome are not of interest. The second is more abstract: "What is the long term average variation in this population?" Here the current generation is of no special interest and data taken from it are used to answer this broader question rather than the earlier, more narrow, one. Theoretical questions, particularly comparisons of the variation to be expected from different selective (or nonselective) mechanisms, are of the second category.

Associated with the estimates of variation which one obtains are standard errors describing the reliability of the estimates. The calculation of standard errors provides problems perhaps more difficult than in the calculation of the estimators themselves. A standard error measures, roughly, the extent of variation in an estimator in different repetitions of the experiment, and it is in the meaning of "repetition" that a distinction must be made concerning the two questions asked in the paragraph above. For the first question, repetition implies taking a different sample from the same DNA segment and in the same population. If, in the generation of interest, the population happens to be extremely homogeneous genetically, the estimated standard deviation will be small. For the second question, repetition can mean various things. It could mean repeating the entire evolutionary process, or repeating the sampling process in different parts of the genome, or (with restriction endonucleases) repeating the sampling process with different recognition sequences. An experimenter might feel, for example, that the particular part of the genome he is examining is of special interest to him and that conceptual sampling of other parts of the genome is not relevant. He might feel that a conceptual re-run of the evolutionary process is similarly not relevant. Nevertheless, standard errors using both of these concepts have been calculated in the literature and for various purposes each are of value. Further, as we see below, the first two ideas of repetition are essentially one and the same, and this observation adds to the value of both concepts.

III. GENETIC MODELS

In this section we consider aspects of the genetic evolution of a monoecious population of fixed size N. (Generalizations to handle dioecy, fluctuations in population size, etc., can be made but do not alter the essence of the development. We thus adhere throughout to the simplest models.)

Consider some gene locus A. We assume this consists of L nucleotide sites, where L is of the order 10^3. There will thus exist the (effectively infinite) number of $M = 4^L$ possible base sequences, or possible "allelic types." For the theory below we consider all M types to be selectively equivalent. Any nucleotide can mutate (with probability v, of order 10^{-9}) to one of the three alternative nucleotide types, with equal mutation rates to all three being assumed. Without significant loss of accuracy we can take the gene mutation rate to be Lv (of order 10^{-6}).

A. Classical Model

The classical model of population genetics, dating to the earliest days and reflecting the laboratory observations of those days, considers two possible alleles, A_1 and A_2, at the locus A. In the molecular framework just discussed this model is accommodated by allowing "A_1" to stand for some defined subset of the M possible allelic types and "A_2" for the remainder and is a useful model if we have a laboratory procedure that can distinguish whether a given gene is in A_1 or A_2, but can provide no further information on its molecular structure.

Suppose in general that the mutation rate from A_1 to A_2 is u_1 and from A_2 to A_1 is u_2. Then the classical Wright-Fisher model of the evolution of the population assumed that if there are i A_1 genes in any given generation, the probability p_{ij} that there will be j in the following generation is

$$p_{ij} = \binom{2N}{j} (\alpha_i)^j (1 - \alpha_i)^{2N-j} ,$$

where $\alpha_i = [i(1 - u_1) + (2N - i)u_2]/(2N)$. A large variety of theoretical conclusions have been derived by assuming this model. The conclusion of most interest to us for our present purposes is that if, at stationarity, a sample of n genes is taken from the population, the probability that it will contain i A_1 genes is

$$\binom{n}{i} \frac{\Gamma(i+b)\Gamma(n-i+a)\Gamma(a+b)}{\Gamma(a)\Gamma(b)\Gamma(n+a+b)} , \tag{1}$$

where $a = 4Nu_1$, $b = 4Nu_2$. It is easily shown for this distribution that

$$P_0 = \text{Prob}(i = 0) = \Gamma(n+a)\Gamma(a+b)/[\Gamma(a)\Gamma(n+a+b)] , \tag{2a}$$

$$P_n = \text{Prob}(i = n) = \Gamma(n+b)\Gamma(a+b)/[\Gamma(b)\Gamma(n+a+b)] , \tag{2b}$$

$$E(i) = nb/(a+b) , \tag{2c}$$

$$E[i(n-i)] = n(n-1)ab/[(a+b)(a+b+1)] , \tag{2d}$$

and that putting $n = 2$, $i = 1$,

$$\text{Prob(heterozygosity)} = 2ab/[(a+b)(a+b+1)] . \tag{3}$$

In the particular case where $u_1 = u_2 = \frac{1}{2}u$, (3) yields

$$\text{Prob(heterozygosity)} = \theta/2(1+\theta) \tag{4}$$

where $\theta = 4Nu$. Thus

$$\text{Prob(homozygosity)} = (2+\theta)/2(1+\theta) \tag{5}$$

Adding the probabilities for $i = 1$ and $i = n$ in (2a) and (2b) we get

$$\text{Prob(sample monomorphism)} = \frac{(1+\frac{1}{2}\theta)(2+\frac{1}{2}\theta)\ \ldots\ (n-1+\frac{1}{2}\theta)}{(1+\theta)(2+\theta)\ \ldots\ (n-1+\theta)} \tag{6}$$

Both (5) and (6) may be regarded as reasonable measures of population homogeneity while (4) is a reasonable measure of population heterogeneity. Estimation of these measures can be regarded as being equivalent to estimation of θ , so far as "long term average" procedures are concerned. We do not pursue this observation here but will do so in some detail for the next model considered.

B. Infinite Alleles Model

This model is appropriate when we have available an apparatus that can tell us whether two genes are, or are not, identical (*i.e.*, do or do not have the same sequence of nucleotides), but can give us no further information (in particular on what the actual sequences are). As noted earlier, the number M of possible allelic types can be taken as infinite and we thus assume that any mutant gene is of an allelic type never before encountered in the population. Each gene mutates with probability u and, as above, we put $\theta = 4Nu$.

A considerable amount of theory is available for this model. We again suppose that, at stationarity, a sample of n $(n << N)$ genes has been taken from the population. If $n = 2$ the probability that the two genes are of different allelic type is

$$\text{Prob(heterozygosity)} = \theta/(1 + \theta) \ , \tag{7}$$

(Kimura and Crow, 1964). Note the difference between this value and that given in (4). More generally, the sample will reveal some (random) number k of different allelic types $(k = 1, 2, \ldots, n)$, with respective numbers $n_1, n_2, \ldots, n_k$ $(\Sigma n_i = n)$. The probability that $k = 1$, so that all alleles observed are of the same allelic type, is (Ewens, 1972)

$$\text{Prob(sample monomorphism)} = (n - 1)!/[(1+\theta)(2+\theta)\ldots(n-1+\theta)] \quad (8)$$

Again, this differs somewhat from the "two-allele" value (6). As with the previous model, measures of heterogeneity and homogeneity are functions of θ, the former increasing with θ and the latter decreasing with θ.

It turns out, in this model, that k is a sufficient statistic for θ, so that once k is given the numbers $n_1, \ldots, n_k$ contain no further information about θ. Standard statistical theory then shows that estimation of any function of θ, in particular the function (7), should be carried out by using k only. The maximum likelihood estimator $\hat{\theta}$ of θ is found by solving

$$k = 1 + \frac{\hat{\theta}}{1+\hat{\theta}} + \frac{\hat{\theta}}{2+\hat{\theta}} + \ldots + \frac{\hat{\theta}}{n-1+\hat{\theta}} \quad (9)$$

and this estimator may be shown to have approximate variance

$$\text{var}(\hat{\theta}) \simeq \theta/\left[\frac{1}{(1+\theta)^2} + \frac{2}{(2+\theta)^2} + \ldots + \frac{n-1}{(n-1+\theta)^2}\right] \quad (10)$$

We return to this variance later when considering restriction endonucleases. From (7) we may estimate the probability of heterozygosity by

$$\hat{\theta}/(1+\hat{\theta}) \quad (11a)$$

More efficiently (Ewens, 1972), the optimal estimator of (7) is

$$\frac{\text{coeff } \theta^{k-2} \text{ in } [(\theta+2)(\theta+3)\ \ldots\ (\theta+n-1)]}{\text{coeff of } \theta^{k} \text{ in } [\theta(\theta+1)(\theta+2)\ \ldots(\theta+n-1)]} \quad (11b)$$

It has been argued that the estimators given by (11a) and (11b) are unreasonable, in particular because they do not depend on $n_1 \ldots n_k$. Thus, for example, (11a) and (11b) give the same estimate of the probability of heterozygosity when $n = 200$, $k = 5$ and

$$n_1 = n_2 = n_3 = n_4 = 1 \text{ , } n_5 = 196 \tag{12}$$

as when

$$n_1 = n_2 = n_3 = n_4 = n_5 = 40 \tag{13}$$

Clearly, it is argued, with the values given in (13) we should estimate a much higher heterozygosity level than with the values (12).

This argument lies at the very heart of the distinction between estimates of long-term averages and estimates of "current-generation" values. If one wished to estimate, from the above data, the long-term frequency (over an extremely large number of generations) with which two genes drawn at random from the population will be of different allelic types, one should use (11a), or, more efficiently, (11b). If on the other hand it is only the present generation that is of interest, it is reasonable to use the estimator

$$1 - \Sigma(n_i/n)^2 \tag{14}$$

which does use the n_i values. We have labored this point because exactly the same question arises in the restriction endonuclease theory we consider later. Estimation of a long-term average is a different matter from estimation of a current generation value. Correspondingly the standard deviation of the estimate of a long-term average can be quite different from that of the current generation value.

C. Infinite Sites Model

In the application of this model we assume an apparatus capable of reading the complete nucleotide sequence of any gene. This apparatus is used to find the sequence in a sample of n genes and from the data so obtained we may calculate the number k^* of "segregating sites in the sample," that is the number of sites where at least two different nucleotide types are observed in the sample. We assume (Watterson, 1975) that there is no recombination between the various sites in the gene. Then with the same assumptions and notations as above,

$$E(k^*) = \theta[1 + \frac{1}{2} + \frac{1}{3} + \ldots + \frac{1}{n-1}] \tag{15}$$

We may use (15) to find an estimator of θ from k^*, namely

$$\hat{\theta} = k^*/[1 + \frac{1}{2} + \frac{1}{3} + \ldots + \frac{1}{n-1}] \tag{16}$$

The variance of this estimator is

$$\mathrm{var}(k^*)/[1 + \frac{1}{2} + \frac{1}{3} + \ldots + \frac{1}{n-1}]^2$$

and Watterson (1975) shows that to a first approximation this is

$$\mathrm{var}(\hat{\theta}) \simeq \theta/\log n\ . \tag{17}$$

In this model the heterozygosity probability (7) still holds and clearly this can be estimated, using (16). The variance of the resulting estimator is approximately

$$\theta/[(1 + \theta)^2 \log n] \tag{18}$$

This variance relates to "long-term average" estimation. If the quantity of interest is the heterozygosity present in the current generation, an estimator similar to (14) would be used, with n_i now meaning the number of nucleotide sequences in the sample of some type "i".

IV. RESTRICTION ENDONUCLEASES

A. Introductory Remarks and Definitions

As in the preceding section, we suppose that the data consist of n homologous DNA segments taken at random from some (diploid) population of size N at stationarity. Each segment is of length L nucleotides, where L is not necessarily known. Each segment has been treated with a restriction enzyme, whose recognition sequence is of length j (j is usually 4 or 6), and the points at which each segment is cut are noted. We make the following definitions and notation:

(i) A *block* is a set of j consecutive nucleotide sites,

(ii) A *cleavage site* is a block for which at least one DNA segment in the sample has the recognition sequence and is therefore cut,

(iii) A block is *monomorphic in the sample* if all members of the sample are genetically identical at that block. (Since restriction fragment data give information only about cleavage sites we will only be able to recognize those blocks monomorphic in the sample that are *monomorphic for the restriction sequence*.) Nonmonomorphic blocks are *polymorphic* in the sample.

(iv) Monomorphism and polymorphism of blocks in the population are defined analogously, although of course from our sample information only we will not know if any block is monomorphic in the population.

(v) The number of cleavage sites in the sample is denoted m and the number of these that are polymorphic in the sample is denoted k ($k = 0, 1, 2, \ldots, m$). We suppose the recognition sequence occurs at c_i members in the sample for polymorphic cleavage site i ($i = 1, 2, \ldots, k$; $c_i = 1, 2, \ldots, n - 1$). The unknown probability of sample nucleotide site polymorphism is p and of population nucleotide site polymorphism is P ($P \geq p$) . In the particular case $n = 2$, the site polymorphism probability p is denoted Q, the site heterozygosity probability.

We have assumed above only one single recognition sequence (of length j). In practice a battery of restriction enzymes will be available. The theory given below can be modified to take such a possibility into account but this does not essentially alter the points made below, and for simplicity we consider only one restriction enzyme.

Our aim is to estimate p, P and Q from the data ($m; k; c_1 \dots c_k$) and to construct standard errors of our estimates. We are thus making inferences from our data to a wider set and, as noted in a previous section, a model (or set of assumptions) is necessary for us to do this. Exactly how detailed this model must be, and what are the minimal and most reasonable assumptions, are not immediately obvious, and we now discuss such models as have been introduced in the literature and the estimators arising from them.

B. Full Genetic Model

In the analysis of Ewens *et al.* (1981) an evolutionary genetic model was introduced to estimate the parameters described above. Consider any particular block in the DNA segment. For each member in the sample the recognition sequence either will, or will not, occur in this block. We identify the recognition sequence with the "allele" A_1 in the model (1) and make

Assumption 1. The population evolution at the block in question is described by the Wright-Fisher model (1), with "A_1" referring to the recognition sequence, "A_2" to all other sequences, with u_1, u_2 defined suitably (see below).

To make use of formulae analogous to (2) we make a symmetry assumption for mutation rates.

Assumption 2 (symmetry). Any nucleotide, of whatever type, mutates with the same probability v, with equal mutation rates to the remaining three nucleotide types.

This assumption, together with the assumption that v is extremely small (of order 10^{-9}) is sufficient to give the mutation rates u_1 and u_2 from the recognition sequence and to the recognition sequence, respectively, as

$$u_1 \simeq jv \tag{19a}$$

$$u_2 = jv(\tfrac{1}{4})^j \tag{19b}$$

They are also sufficient to calculate the parameters p , P and Q defined earlier. We find (Ewens *et al.*, 1981), if n is not small,

$$Q \simeq \xi \tag{20}$$

$$p \simeq \xi \log n \tag{21}$$

$$\begin{aligned} P &\simeq 1 - (2N)^{-\xi} \\ &\simeq \xi \log 2N \quad \text{for very small } \xi \end{aligned} \tag{22}$$

where we define ξ by $\xi = 4Nv$. Similarly, we find

$$\text{Prob(sample block polymorphism)} \simeq j\,\xi \log n\ , \tag{23}$$

$$\text{Prob(conditional sample block polymorphism)} \simeq 2j\,\xi \log n \tag{24}$$

It is necessary to define the probability (24). This is the probability, given at least one recognition sequence in the sample at the block in question, that the sample is not monomorphic for the recognition sequence.

The assumptions made above, in particular that of the small value of v , imply two further assumptions which strictly need not be stated, but which will be brought out here to assist with comparison with the analyses described later. These are:

<u>Assumption 3</u>. If polymorphism exists in the sample in the various sequences at any block, this polymorphism is due to a polymorphism at one nucleotide site only.

<u>Assumption 4</u>. If polymorphism exists in the sample at a given nucleotide site, exactly two base types are segregating at that site.

Recall now that k of the m cleavage sites in the sample are polymorphic. It is then tempting, using (22), to estimate $j \xi \log n$ by k/m and from (21) to estimate p by k/jm. This procedure has been used in the literature (see, for example, Jeffreys, 1979). This intuitive estimator is, however, biased. We know, from the definition of a cleavage site, that the recognition sequence occurs there at least once and therefore k/m is an estimator not of (23) but of (24). We thus derive, from this analysis, the estimator $\hat{p}_1$ of p, defined by

$$\hat{p}_1 = k/(2jm) \tag{25}$$

We may similarly estimate Q by

$$\hat{Q}_1 = 1/(2j\ m \log n)$$

and, if N is known or can be reasonably estimated, estimate P by

$$\hat{P}_1 = (k \log 2N)/(2j\ m \log n) \tag{27}$$

Can the error (by a factor of 2) in the previously used intuitive estimator of p be reached, and reasonable estimators of p found, by using fewer, more reasonable, or nongenetically based, assumptions? We now turn to analyses that suggest the answer is "yes".

C. Engels' Estimator

Engels (1981) produced an estimator of p using a beguilingly simple assumption, namely

Assumption 5. The frequency of the recognition sequence among blocks monomorphic in the sample has the same mean value as its frequency in the sample as a whole, or, in terms of the notation introduced earlier,

$$E(m - k)/L = \text{Prob(sample block monomorphism)} \times \text{Prob(recognition sequence occurs in any given block)} \tag{28}$$

We know that the recognition sequence occurs $n(m - k) + \sum_{i=1}^{k} c_i$ times in our sample, so we may estimate the second term on the right-hand side by

$$[n(m - k) + \Sigma c_i]/nL$$

This leads to the estimate

$$n(m - k)/[n(m - k) + \Sigma c_i]$$

for the probability of sample block monomorphism. Using now also Assumption 3, we get from this simple but ingenious argument a second estimator p_2 of p , namely

$$\hat{p}_2 = (\Sigma c_i)/\{j[n(m - k) + \Sigma c_i]\} \tag{29}$$

This estimator, unlike (25), uses the values $c_1 \ldots c_k$ and appears initially to differ considerably from (25). If, however, we were to add the "symmetry" assumption 4, it would be reasonable to replace, in (29), the random variables $c_1 \ldots c_k$ by n/2 , and if we do this, $\hat{p}_2$ becomes

$$\hat{p}_3 = k/[j(2m - k)] \tag{30}$$

which, if k is small compared to m (as often occurs in practice), does not differ substantially from (25). Thus the advantage of (29) over (25) appears to be that it requires neither the symmetry assumption 4 nor the machinery of the evolutionary Wright-Fisher genetic model. To discuss this view we subject Assumption 5 to a closer scrutiny.

Suppose we play a large collection of games, each game consisting of n tosses of a coin. If the probability of head on each toss is p, the frequency of heads in the entire set of tosses has mean value p. Consider now only those games where all n tosses give the same result (*i.e.*, all heads or all tails). The mean value of the fraction of such games in which all n tosses give heads is $p^n/[p^n + (1-p)^n]$. This will equal p only in the symmetric case $p = 1/2$ (leaving aside the trivial cases $p = 0$, $p = 1$). Suppose now we replace, in Assumption 5, "recognition sequence" by "heads" and "blocks monomorphic in the sample" by "games where all n tosses give the same result." Then Assumption 5 would be correct for the coin-tossing game only if the coin is fair. While the coin-tossing game is not an appropriate analog of the genetic process, our discussion shows that the assumption is a deeper one than first appears and must be justified, in the genetic context, by further argument (or shown to be reasonably acceptable by observation). If we think of Assumption 5 as implying some class of genetic models, this is tantamount to justifying that class to be appropriate for the restriction endonuclease data.

D. Hudson's Estimator

Hudson (1982) derived a third estimator of p by starting with the probability theory identity

$$\text{Prob}(A|B) = \text{Prob}(A)\ \text{Prob}\ (B|A)/\text{Prob}(B) \tag{31}$$

where A = "sample block polymorphism"

B = "block is a cleavage site"

Let the probability that a block monomorphic in the sample is a cleavage site be r. By making Assumptions 2, 3 and 4 above, we find immediately

$$\text{Prob}(B|A) = \text{Prob(block polymorphic in sample is a cleavage site)} = 2r$$

This follows immediately since a polymorphic block contains exactly two different sequences, giving two "chances" for the recognition sequence. Symmetry is invoked to give all possible sequences equal probabilities. Further, from Assumption 3, Prob(A) = jp. The numerator on the right-hand side of (31) thus becomes $2rjp$.

The denominator is the sum of this value and the probability that a cleavage site is monomorphic in the sample, and the assumptions made show immediately that this is

$$2rjp + r(1 - jp)$$

The right hand side in (31) is thus

$$2jp/(1 + jp)$$

We may now estimate p by equating this ratio to the observed fraction k/m to get

$$\hat{p}_3 = k/[j(2m - k)]$$

which is identical to (30). This is as we expect since (30) is found from (29) by invoking the "symmetry" assumption 4 which is already used in the above derivation.

We turn now to estimates of site heterozygosity and first consider the special case $n = 2$. Here, automatically, $\Sigma c_i = k$ and (29) becomes (30). In samples of two DNA segments, p is identically the site heterozygosity probability and so the latter is estimated, under both Engels' and Hudson's approach, by (30). In estimating the heterozygosity probability from samples larger than two Engels arrives, by an argument parallel to that leading to (29), at the estimator

$$\hat{Q}_2 = (n\Sigma c_i - \Sigma c_i^{\,2})/\{j(n - 1)[(m - k)\, n + \Sigma c_i]\} \tag{32}$$

At first sight this may seem to differ considerably from the genetically-derived expression (26). The two are, however, quite close. To show that this is so we calculate the expected value of $c_i(n - c_i)$ and c_i in the genetic model (1), recalling that we have here the added restriction $1 \leq c_i \leq n - 1$. Thus, using (2a), (2b), (2c) and (19) we get

$$a = 4Njv = j\xi \ , \ b = 4Njv\ (\tfrac{1}{4})^j = j\xi(\tfrac{1}{4})^j \ ,$$

$$E(c_i) = [nb/(a + b) - nP_n]/[1 - P_0 - P_n]$$

$$Ec_i(n - c_i) = n(n - 1)ab/[a + b)(a + b + a)(a - P_0 - P_n)] \tag{33}$$

Using elementary asymptotic theory for the gamma function, these become, approximately,

$$E(c_i) \simeq n/2 \qquad \text{(as we expect)} \tag{34}$$

$$Ec_i(n-c_i) \simeq n(n - 1)/2 \log n \tag{35}$$

Recalling that the numerator in (32) consists of k terms we get, by replacing all expressions in (32) by their expected values in the genetic model,

$$k/[j(2m - k) \log n] \tag{36}$$

This is extremely close to (26) and differs from (26) in precisely the same way that (25) differs from (30). Thus on average (32) and (26) will be close, assuming the Wright-Fisher genetic model to hold.

The most useful comparison of (32) and (26) relates to the comparison of "current-generation" and "long term" estimation. In the genetic model, if we require (using restriction endonuclease data) an estimate of the heterozygosity in the current generation, we might prefer to use (32). On the other hand, if we seek a long-term average we would use (26) [or perhaps (36)]. The arguments in favor of these two approaches are identical to the arguments in favor of using (14) for current-generation estimation and (11) for long-term estimation of heterozygosity in the infinite alleles model considered earlier.

Thus the differences between the estimates $\hat{p}_1$, $\hat{p}_2$ and $\hat{p}_3$ and $\hat{Q}_1$ and $\hat{Q}_2$ are two-fold. First, each estimate is arrived at by making a different set of assumptions and in practice which estimator is used will depend on which set of assumptions, or which model, one is prepared to accept. Secondly, we can regard $\hat{p}_2$ and $\hat{Q}_2$ as focusing on the current generation and $\hat{p}_1$, $\hat{p}_3$ and $\hat{Q}_1$ on long-term averages, so that which estimator is used will also depend on whether "current-generation" or "long-term average" values are of interest. This latter point is particularly relevant when we turn to the standard deviations of the estimates.

V. STANDARD DEVIATIONS

We wish to calculate the standard deviations of the estimators $\hat{p}_1$, $\hat{p}_2$ and $\hat{p}_3$. A standard deviation measures, roughly, the expected variability of an estimator in a sequence of repetitions, and the problem of the concept of repetitions, and whether the long-term or current-generation value of p is being estimated, immediately arises.

Before considering the calculations we take up a minor technical point. An estimator such as (25) is not strictly well defined, since it is possible that $k = m = 0$. If this happens the observer has no cleavage sites and thus no data and would hardly be considering any estimation procedure. We thus assume that the segment of DNA examined is so large that problems of this nature do not arise. In general we will use large-sample theory and various approximations that assume m is large in the calculations we give below.

We consider first repetitions of the entire evolutionary process, so that we calculate standard errors analogous to (10) and (18). Take first the estimator $\hat{p}_1$, defined in (24). Given m, we may reasonably assume k has a binomial distribution with index m and parameter $2j\,\xi \log n$ [see (24)]. This gives $E(\hat{p}_1) = p$, as desired. The variance of $\hat{p}_1$ can then be calculated as

$$E_m E^*(\hat{p}_1^2) - p^2$$

where E^* denotes expected values given m, and E_m denote expectation of any function of m. We find, if terms of order ξ^2 are ignored, that

$$\text{var}(\hat{p}_1) \simeq \xi \log n/[2jE(m)] \tag{37}$$

Using the sample data we would estimate this variance by

$$\hat{p}_1/(2jm) = \hat{p}_1^2/k \tag{38}$$

The estimator $\hat{p}_3$ does not differ much, for small k, from $\hat{p}_1$ and its variance estimate will be similar to (38), *i.e.*

$$\text{var}\ \hat{p}_3 \simeq \hat{p}_3^2/k \tag{39}$$

The variance of $\hat{p}_2$ is rather harder to approximate. Under the assumption that $E(c_i) = n/2$, Hudson (1982) uses the theory of statistical differentials to find

$$\text{var } \hat{p}_2 \simeq \hat{p}_3^2 \; [1 + 4 \text{ var}(c_i)/n^2]/k \tag{40}$$

This is clearly larger than (39), but the form of (40) indicates exactly why this is so. The estimator $\hat{p}_3$ can be found from $\hat{p}_2$ by replacing each c_i by n/2 , and such "c_i" values have zero variance. Then (40) reduces immediately to (30). For practical purposes, (40) can be estimated by estimating var(c_i) by

$$\Sigma(c_i - n/2)^2/k \tag{41}$$

It is of interest to note that in the Wright-Fisher genetic model, (34) and (35) give

$$\text{var}(c_i) = n^2/4 - n(n - 1)/(2 \log n)$$

so that in this case, if n is not small,

$$\text{var}(\hat{p}_2) \simeq \text{var } \hat{p}_1 \; [2 - 2/\log n]$$

We emphasize that the fact that var($\hat{p}_1$) and var($\hat{p}_3$) are less than var($\hat{p}_2$) does not follow from any intrinsic superiority of the former two, but rather because the former two are calculated making an additional assumption (Assumption 4) not made in arriving at $\hat{p}_2$.

The estimator $\hat{Q}_1$ is identically equal to $\hat{p}_1/\log n$ and hence from (37),

$$\text{var}(\hat{Q}_1) = \xi/[2j \log n \; E(m)] \tag{42}$$

This may be estimated by $\hat{Q}_1^2/k$.

It was noted in (20) that Q and ξ are essentially identical, so that we may regard $\hat{Q}_1$ as an estimator of ξ rather than Q . Taking this view, it is interesting to compare the variance formula (42) with parallel formulae for the infinite alleles and infinite sites models. To do this we must suppose, first, that the DNA segment treated by restriction enzymes is exactly one gene, comprising L nucleotide sites. Now ξ is defined in terms of nucleotide mutation rates [see immediately below equation (22)] whereas the parameter θ [introduced below equation (4)] is defined from gene mutation rates. We should therefore put $\theta = L\xi$ and estimate θ by $L\hat{Q}_1$. The variance of this estimator, from (42), is approximately

$$\mathrm{var}(\hat{\theta}) \simeq L^2\xi/[2j \log n \; E(m)]$$

It may be shown that $E(m) \simeq (\frac{1}{4})^j L\,(1 + \xi j \log n)$ and thus, if $\xi j \log n$ is small,

$$\mathrm{var}(\hat{\theta}) \simeq \theta 4^j/(2j \log n) \tag{43}$$

It is of great interest to compare this with (10) and (17). Clearly (43) is a much larger value than the other two estimates and this reflects the fact that the restriction endonuclease technique provides information on a very limited part of the DNA considered, while the estimators leading to (10) and (17) use the entire DNA segment. This illustrates the value of computing "long-term average estimate" variances for comparing the efficiencies of various laboratory techniques for investigating genetic variation.

In calculating his estimate of $\mathrm{var}(\hat{p}_3)$, Hudson (1982) did not refer to conceptual re-runs of the evolutionary process, but rather to repetition where either (i) different restriction sequences are used for the given n DNA segments or (ii) different parts of the genome are sampled. The two are equivalent mathematically and since the latter is simpler conceptually we now consider that one.

If random segments of the genome (each of length L nucleotides) are sampled, k and $m - k$ are independent random variables which, to a close approximation, can be taken as having Poisson distributions with parameters

$$2j(\tfrac{1}{4})^j L\ \xi \log n \quad \text{and} \quad (\tfrac{1}{4})^j L\ (1 - \xi j - \log n)$$

respectively. [This confirms the expected value of m given before equation (43).] Writing

$$\hat{p}_3 = k/\{j[2(m - k) + k]\}$$

and using the theory of statistical differentials, the variance of $\hat{p}_3$ becomes, approximately,

$$\xi \log n/[2(\tfrac{1}{4})^j Lj] \tag{44}$$

and this is estimated from the data by $\hat{p}_3^2/k$. *This is exactly the same as the value found from (39).* In other words, variance from one conceptual repetition of the evolutionary process to another, at the same DNA segment, is the same as variance in one repetition of the evolutionary process from one DNA segment to another. This follows directly from the argument that since the different segments can be taken as evolving independently, one is in fact sampling repetitions of the evolutionary process by moving to different parts of the genome. It follows immediately that under the concept of repetition by moving from one part of the genome to another, the variance of $\hat{p}_2$ is given, approximately, by (40), with $\text{var}(c_i)$ estimated as in (41).

Consider now variances relating to conceptual resampling the same DNA segment, in the same population in the same generation, but with different individuals in the sample. Here we label blocks $1, 2, \ldots, \ell, \ldots, L$ and suppose that the frequency of the recognition sequence at block ℓ is, in the population, π_ℓ . The number of times the recognition sequence occurs in a sample of n at this block can be taken as binomial with index n , parameter π_ℓ . Using the theory of statistical differentials, Engels (1981) finds from (29) that the variance of $\hat{p}_2$ is approximately

$$\mathrm{var}(\hat{p}_2) = \left(\frac{n}{jC}\right)^2 \sum_{\ell} \{\pi_\ell^n(1 - \pi_\ell^n)$$

$$+ \frac{n(\bar{m}-\bar{k})\pi_\ell(1-\pi_\ell)}{C}\left[\frac{\bar{m}-\bar{k}}{C} - 2\pi_\ell^{n-1}\right]\} \qquad (45)$$

Here $C = E[n(m-k) + \Sigma c_i]$, $\bar{m} = E(m)$, $\bar{k} = E(k)$. In the estimation of this variance we replace C , $\bar{m}$, $\bar{k}$ and π_ℓ by their estimates $n(m - k) + \Sigma c_i$, m , k and c_i/n . The sum in (45) then contains positive terms only for polymorphic cleavage sites.

It is of great interest to compare (40) and (45) and this must be done numerically. We provide examples using numerical values typical of those occurring in practice.

Example 1. Suppose n , the number of homologous DNA segments examined, is 200, the length j of the recognition sequence is 6, the number m of cleavage sites is 100 and that $k = 10$ of these are polymorphic in the sample. At these ten polymorphic sites, suppose that c_i takes the value 10 at five and the value 190 at the remaining five. The estimator $\hat{p}_2$ of sample site polymorphism is, from (29),

$$1000/[6(18000 + 1000)] = .00877 \qquad (46)$$

[The estimator p_3 , calculated from (29), takes exactly the same value.] Using (39) and (40) we estimate the variance of this estimator by

$$(.00877)^2(1 + .81)/10 = 1.39 \times 10^{-5} \qquad (47)$$

when the concept of repetition refers to sampling different parts of the genome (or alternatively to different conceptual re-runs of the evolutionary process). Suppose now that the concept of repetition refers to sampling different individuals in the population, using the same DNA segment, so that (45) is appropriate. The variance estimate becomes

$$7.05 \times 10^{-9} \tag{48}$$

This is less than (47) by a factor of almost 2000. Clearly, the concept of repetition involved in calculating a variance (and hence a standard error) of an estimator plays a crucial part in assessing the accuracy of that estimator.

It is easy to see why this is so. If, at a polymorphic cleavage site, the value of c_i is not close to 0 or 200, it is almost certain that sampling different members of the population would also reveal a polymorphic cleavage site. Similarly, a monomorphic cleavage site would quite possibly be a monomorphic cleavage site with a different sample. No such claims can be made when a different part of the genome is sampled and this leads to a much larger variance under this concept of repetition than under resampling the population using the same DNA segment.

The case where the two variances will be closest is when at polymorphic cleavage sites the frequency of the recognition sequence is either extremely high or extremely low. We illustrate this by

<u>Example 2</u>. Suppose n, m and k are as in Example 1, but that, at the ten polymorphic cleavage sites, c_i takes the value 1 at five and 199 at the remaining five. The variance estimate (28) is

$$1.52 \times 10^{-5} \tag{49}$$

while (43) yields

$$5.59 \times 10^{-6} \tag{50}$$

The ratio of these is 2.72 and this is not a coincidence: it may be shown that the asymptotic value of this ratio is $e(= 2.718)$. Here there is considerable variance in the estimate of p by re-sampling the population at the same DNA segment since the data suggest that some sites polymorphic in one sample will be monomorphic in another (and vice versa), and that some cleavage sites in one sample will not be cleavage sites in another. This approaches the situation in sampling from different parts of the genome.

We have just discussed the variance of $\hat{p}_2$ when repetition refers to conceptual re-sampling the current generation of the population, using the same part of the genome, and compared the variance found [*viz.* (45)] with that applying when repetition means either conceptual re-runs of the revolutionary process or, equivalently, sampling different parts of the genome.

We turn now to the variance of $\hat{p}_1$ and $\hat{p}_3$ when repetition refers to re-sampling the current generation, using the same part of the genome. The estimators $\hat{p}_1$ and $\hat{p}_3$ are quite similar and since $\hat{p}_1$ is of a slightly simpler algebraic form, we consider its variance rather than that of $\hat{p}_3$.

Assuming m is large and that the main part of the variance in $\hat{p}_1$ comes from variation in k, we have

$$\text{var}(\hat{p}_1) = \text{var}(k)/(2mj)^2 \tag{51}$$

under this form of repetition. The value of k in a new sample can differ from the value in the actual sample for four different reasons:-

(1) a polymorphic cleavage site in the actual sample can be a monomorphic cleavage site in the new sample or

(2) cease to be a cleavage site in the new sample;

(3) a monomorphic cleavage site in the actual sample can be a polymorphic cleavage site in the new sample, or

(4) a noncleavage site in the actual sample can be a polymorphic cleavage site in the new sample.

For the full genetic model, embodying Assumptions 1 and 2, the probability of each of these events can be calculated. For example, the probability of (1) is, using a convenient shorthand,

$$\text{Prob(actual poly, new mono)/Prob(actual poly)} \\ = [\text{Prob(new mono)} - \text{Prob(new mono, actual mono)}]/ \\ \text{Prob(actual poly)} \tag{52}$$

These probabilities can be calculated from equations (2a) and (2b), using (19) to give a and b, and noting that Prob(new mono, actual mono) is the same as Prob(mono cleavage site in a sample of $2n$). Thus, from (23) and (24) the expression (52) becomes, approximately,

$$[(1 - j\xi \log n) - (1 - j\xi \log 2n)]/2j\xi \log n \\ = \log 2/(2 \log n) \tag{53}$$

The probabilities for (2), (3) and (4) are similarly found to be, respectively,

$$\log 2/(2 \log n) \ , \tag{54}$$

$$j\xi \log 2 \ , \tag{55}$$

$$j\xi \left(\tfrac{1}{4}\right)^j \log 2 \tag{56}$$

Note that these probabilities satisfy "balancing" requirements: for example, noting that $E(k/m) = 2j\xi \log n$, the mean numbers of events of types (1) and (3) are equal, and similarly with (2) and (4).

The difference between the values of k in the actual and the new sample can be written as

$$\delta(k) = -\delta_1 + \delta_2 - \delta_3 + \delta_4$$

where δ_i is the number of events of type i $(i = 1, 2, 3, 4)$. Given m, the means of each of the δ_i are approximately $mj\xi \log 2$. If we assume independent Poisson distributions for each of these, we get

$$\mathrm{var}(\delta) \simeq 4\ mj\ \xi \log 2 \tag{57}$$

This is $2\ \mathrm{var}(k)$ and so we get from (51), as our estimate of the variance of $\hat{p}_1$ under this concept of repetition,

$$(\hat{p}_1)^2 \log 2/(k \log n) \tag{58}$$

This variance bears a simple relationship to the value given in (38) for repetition by sampling different parts of the genome. The larger the sample size the smaller is the variance (58), but, in contrast to classical statistics when variances of averages decrease as n^{-1}, here the variance decreases only as $(\log n)^{-1}$ This occurs, of course, because of the high degree of dependence in the genetic make-up of the population in any generation from one individual to another.

How does the variance (58) compare with (45)? For the numerical values given in Example 1, var $(\hat{p}_1)$, calculated from (58), is

$$9.085 \times 10^{-7} \tag{59}$$

This differs from (48) by a factor of about 80 and this is presumably due to the increased efficiency, in estimating current generation values, in using an estimator that employs the c_i values rather than an estimator $\hat{p}$ (or $\hat{p}_3$) that does not. Note, however, that for the numerical values given in Example 2, (59) remains unchanged and this is now less than the numerical value (50). Why this is so is not clear.

The aim of the experimenter is of course relevant to which concept of repetition is appropriate. If the particular DNA segment is of importance to him, the concept of repetition by sampling different parts of the genome is presumably not of interest: he is concerned only with one particular part of the genome. On the other hand, if he is comparing, broadly, the genetic variability in one species to that in another, repetition by sampling different parts of the genome is possibly appropriate. Further, as we have seen in comparing (43) with (10) and (17), this latter concept of repetition is relevant when comparing the efficiency of one technique over another in estimating genetic variation.

VI. POPULATION PARAMETERS

Most of the above discussion refers to estimation of the sample parameter p . Often, however, neither the particular sample taken, nor indeed any sample of the same size, is really of central interest, and one is concerned, rather, with the population parameter P . We now turn to estimates of this quantity.

In the full genetic model, embodying Assumptions 1 and 2 (and implicitly 3 and 4), P is given by (22) and is estimated by (27). This requires an extrinsic estimate of N (which need not be too accurate as N occurs only in a logarithm). Since

$$\hat{P}_1 = \hat{p}_1 \,(\log 2N/\log n)$$

the variance of this estimate would be estimated from (38) as

$$\hat{p}_1^2 \,(\log 2N)^2/[k(\log n)^2] \tag{60}$$

when repetition implies either conceptual re-runs of the evolutionary process or using different parts of the genome, or [see (58)] by

$$(\hat{p}_1)^2 \log 2 \,(\log 2N)^2/[k(\log n)^3] \tag{61}$$

when repetition implies re-sampling the current population. It is not clear how to generalize (29) and (45) to obtain an estimate of P , and a variance estimator of this estimate, using $\hat{p}_2$ as a starting point.

VII. CONCLUSIONS

There is no single correct model for restriction fragment data generation and hence no single correct estimator of various sample and population characteristics and in particular of the degree of genetic diversity in the population. Different models, and hence different estimators, have been proposed in the literature and are discussed here.

Further, there is no unique concept of the standard deviation of an estimator. Standard deviations refer to the degree of variation of an estimator under repetition of the sampling procedure, but different concepts of repetition are possible and lead to different standard errors. Care is needed in comparing different estimators by calculation of their various standard errors, since different conventions may be used for each. Further, a small standard error does not necessarily imply a good estimator: it may arise from making restrictive assumptions not made in forming another estimator. The assumptions behind each estimator and a correct comparison of standard errors must be examined before the relative merits of two estimators can be assessed.

VIII. ACKNOWLEDGMENTS

A large number of results given in this chapter were first obtained by W. R. Engels or R. R. Hudson. It is a pleasure to acknowledge my indebtedness to both of them for many stimulating discussions and considerable guidance on the topics discussed above.

Chapter 4

STATISTICAL ANALYSIS OF RESTRICTION ENZYME MAP DATA AND NUCLEOTIDE SEQUENCE DATA

Norman Kaplan

Biometry and Risk Assessment Program, NIEHS, Research Triangle Park, North Carolina

I. INTRODUCTION

In recent years restriction enzymes and DNA cloning and sequencing techniques have been used successfully to detect DNA differences within and between populations. These techniques are particularly useful since they can be used to isolate and study sequences from the noncoding as well as the coding portions of the genome.

A restriction enzyme generally cleaves the DNA at a unique sequence of nucleotides, usually four or six base pairs long. Some restriction enzymes, however, cleave the DNA at several different sequences. If any of these restriction enzymes are used to study the DNA, then minor modifications of some of the theory in this chapter must be made. Hence, to simplify the development, it will be assumed here, unless stated otherwise, that the restriction enzymes recognize unique sequences. The location of a recognition sequence will be called a *cleavage site*, and the term *site* will be used for any consecutive sequence of nucleotides of the same length as the recognition sequence.

By using several different restriction enzymes, it is possible to construct a "map" of the DNA sequence under study which specifies the relative locations of the cleavage sites. The maps of homologous DNA sequences can then be compared to detect differences. The homologous sequences could have come from a variety of sources, *e.g.*, individuals within a population, individuals from different species, or a region within a single genome which underwent gene duplication and then diverged. This technique has, for example, been used to study mitochondrial DNA (mtDNA) sequence variation within and between different populations (Brown and Vinograd, 1974; Upholt and Dawid, 1977; Brown *et al.*, 1979; Shah and Langley, 1979; Brown, 1980). The resulting data have provided important information on the rate of nucleotide substitution and the genetic structure of populations.

Restriction enzyme map data provide only a limited amount of information about the similarities of homologous DNA sequences. Complete information is in hand when the DNA is totally sequenced. Recent developments in DNA sequencing techniques (Maxam and Gilbert, 1977; Sanger *et al.*, 1977) allow a rapid determination of the DNA sequence, and so this approach is becoming more practical. For example, the mtDNA sequences of man (Anderson *et al.*, 1981), cow (Anderson *et al.*, 1982) and mouse (Bibb *et al.*, 1981) are now known.

The analysis of the data generated by these new techniques has required the development of new statistical methodology. Several authors have used this data to estimate evolutionary distance between two DNA sequences having a common ancestor (Upholt, 1977; Gotoh *et al.*, 1979; Kaplan and Langley, 1979; Nei and Li, 1979; Kimura, 1980, 1981; Aoki *et al.*, 1981; Kaplan and Risko, 1981, 1982; Takahata and Kimura, 1981). Others have used the data to estimate DNA sequence variation within populations (Engels, 1981; Ewens *et al.*, 1981; Nei and Tajima, 1981; Hudson, 1982). The purpose of this chapter is to survey these statistical methods.

In the next section several different estimates of evolutionary distance are reviewed. All of these estimates are constructed within the framework of a detailed mathematical model for the substitutional process at the nucleotide level. Hence, a careful development of this model is presented. Estimates of polymorphism within a population are discussed in the third section of the chapter. These estimates are less model-dependent (see discussion in the chapter by Ewens in this volume), although certain key mathematical assumptions are required to make the analysis tractable. In the final section of the chapter, some simulation results will be presented which allow a comparison of the performance of the various estimates. Some data are also analyzed to illustrate the methods.

II. ESTIMATING EVOLUTIONARY DISTANCE

A. The Model

One use of restriction enzyme map data and nucleotide sequence data has been to estimate evolutionary distances between two homologous DNA sequences that are assumed to have diverged from a common ancestor τ years ago. Most of the estimates that have been proposed in the literature have been derived within the context of an underlying mathematical model which describes the evolutionary process at the nucleotide level. Each of these models is a special case of the following general model.

Suppose there are M nucleotides in the sequence of interest, and they are numbered in order starting from a predetermined origin. For linear DNA, the origin is usually the 3' or 5' end, while for circular DNA the origin is arbitrary. Each of the M nucleotides is said to be in state 1, 2, 3 or 4 if it is a G, C, A, or T. For one of the sequences let $Y_i(t)$ denote the state of the *ith* nucleotide at time t, where time is measured from the point of divergence. Thus, $Y_i(0)$ denotes the state at divergence and $Y_i(\tau)$ the current one. To simplify notation, τ will be suppressed and $Y_i(\tau)$ will be written as Y_i.

The process $[Y_i(t),\ 0 \leq t \leq \tau]$ is assumed to be stochastic and should be thought of as describing the evolutionarily accepted mutations of the *ith* nucleotide. There are many different causes of mutational change in DNA. Some possible ones are base pair substitutions, deletions of base pairs, and insertions of base pairs. In this chapter it will be assumed for simplicity that mutation arises only from base pair substitutions.

It is further assumed that all the nucleotides evolve in the same way, *i.e.*, the $[Y_i(\cdot)]_{i=1,\ldots,M}$ are identically distributed processes. In certain cases this assumption must be modified, *e.g.*, for DNA from a coding region it is more appropriate to study the first, second, and third position nucleotides separately. Other authors also assume that the M processes are stochastically independent and so this assumption will be made here as well.

The simplest stochastic structure which can be assumed for the Y process is that of a four-state continuous time Markov chain (Karlin and Taylor, 1975). The following is a convenient description of this process. Suppose that whenever a nucleotide is in state j (j = 1,2,3,4), it remains in this state for a random amount of time T_j, where the distribution of T_j is given by

$$P(T_j < t) = 1 - e^{-\lambda(j)t}, \qquad t > 0$$

The parameter $\lambda(j)$ should be interpreted as the rate at which evolutionarily acceptable mutations occur when a nucleotide is in state j. Whenever an evolutionarily acceptable mutation occurs, the state changes to k with probability p_{jk}. Let

$$P = (p_{jk})_{1 \leq j,k \leq 4} \qquad \text{and} \qquad \underline{\lambda} = [\lambda(1), \lambda(2), \lambda(3), \lambda(4)]$$

For the present, the only restriction that is placed on the process is that it be stationary, *i.e.*, if $p_j(t) = \text{Prob}[Y(t)=j]$, then $\underline{p}(t) = [p_1(t),\ p_2(t),\ p_3(t),\ p_4(t)]$ does not depend on t. Let $\underline{p} = (p_1,\ p_2,\ p_3,\ p_4)$ denote the stationary probabilities. Then it is necessarily true that

$$\underline{p}P = \underline{p} \tag{1}$$

At this point it is constructive to consider some of the choices for P and $\underline{\lambda}$ that have been proposed.

Example 1. The simplest case is that used by Jukes and Cantor (1969), namely

$$P = \begin{bmatrix} 0 & 1/3 & 1/3 & 1/3 \\ 1/3 & 0 & 1/3 & 1/3 \\ 1/3 & 1/3 & 0 & 1/3 \\ 1/3 & 1/3 & 1/3 & 0 \end{bmatrix}, \quad \underline{\lambda} = (\lambda,\lambda,\lambda,\lambda)$$

The stationary probabilities associated with this matrix are all equal to 1/4. This choice for P and $\underline{\lambda}$ was implicitly made by Upholt (1977), Gotoh *et al.* (1979), and Nei and Li (1979). The model considered by Kimura (1980) has all the $[\lambda(i)]$ equal, and equal stationary probabilities, but has a different P matrix.

One quantity at the molecular level that can easily be measured is the GC content of the sequence. It has become clear that, for mtDNA, GC content is not always equal to 1/2 (Anderson *et al.*, 1981). This prompted Kaplan and Langley (1979) to modify Example 1 in the following way.

Example 2. Let α = GC content. Then:

$$P = \begin{bmatrix} 0 & \alpha & (1-\alpha)/2 & (1-\alpha)/2 \\ \alpha & 0 & (1-\alpha)/2 & (1-\alpha)/2 \\ \alpha/2 & \alpha/2 & 0 & 1-\alpha \\ \alpha/2 & \alpha/2 & 1-\alpha & 0 \end{bmatrix} \quad \underline{\lambda} = (\lambda,\lambda,\lambda,\lambda)$$

The stationary probabilities for this matrix are

$\underline{p} = [\alpha/2, \alpha/2, (1-\alpha)/2, (1-\alpha)/2]$

Recently, Aoki *et al.* (1981) and Takahata and Kimura (1981) have considered models where the distributions of the waiting times are state dependent, *i.e.*, not all the $\lambda(i)$ are equal. One of the simpler examples of this type of model is the following:

Example 3. (Takahata and Kimura)

$$P = \begin{bmatrix} 0 & \gamma_1 & \beta' & \varepsilon' \\ \gamma_1 & 0 & \varepsilon' & \beta' \\ \alpha' & \delta' & 0 & \gamma_2 \\ \delta' & \alpha' & \gamma_2 & 0 \end{bmatrix} \qquad \underline{\lambda} = [\lambda(1), \lambda(1), \lambda(2), \lambda(2)]$$

where

$$\gamma_1 = \frac{\gamma}{\lambda(1)}, \quad \beta' = \frac{\beta}{\lambda(1)}, \quad \varepsilon' = \frac{\varepsilon}{\lambda(1)}, \quad \lambda(1) = \gamma + \beta + \varepsilon$$

and

$$\gamma_2 = \frac{\gamma}{\lambda(2)}, \quad \alpha' = \frac{\alpha}{\lambda(2)}, \quad \delta' = \frac{\delta}{\lambda(2)}, \quad \lambda(2) = \gamma + \alpha + \delta$$

Thus, five parameters need to be specified. Aoki *et al.* (1981) also proposed a model which has variable $\lambda(i)$ and a P matrix that allows for a GC content different from 1/2.

B. Evolutionary Distance

Let

$$\bar{\lambda} = \sum_{i=1}^{4} p_i \lambda(i)$$

and set

$$K = 2\bar{\lambda}\tau$$

The quantity K is defined to be the measure of evolutionary distance between two homologous DNA sequences. It is interpreted as the average number of evolutionarily accepted substitutions per nucleotide that have occurred since divergence.

The estimation of K from restriction enzyme map data or nucleotide sequence data is too difficult in general since K depends on all the $\lambda(i)$. However, for special cases progress has been made. The simplest case is when all the $\lambda(i)$ are equal to, say, λ and hence $K = 2\lambda\tau$. Estimates of K for certain special cases when the $\lambda(i)$ are not all equal have also been developed by Aoki *et al*. (1981) for restriction map data, and Takahata and Kimura (1981) for nucleotide sequence data. Since these estimates of K are complicated and hold only for special cases, they will not be discussed in this chapter. The reader is referred to the above papers for details. For the remainder of the chapter it will therefore be assumed that all the $\lambda(i)$ are equal and this common value will be denoted by λ.

Before proceeding to the estimates of K or, equivalently, of $\lambda\tau$, it is convenient to define the following quantities. Let

P_{ij} = probability that a nucleotide in one of the sequences is currently in state j given that it was in state i at time zero

When all the $\lambda(i)$ are equal to λ, it is not difficult to show that

$$P_{ij} = e^{-\lambda\tau} \sum_{\ell=0}^{\infty} \frac{(\lambda\tau)^{\ell}}{\ell!} p_{ij}^{(\ell)} \tag{2}$$

where $p_{ij}^{(\ell)}$ is the ij*th* entry of $P^{(\ell)}$ and $P^{(\ell)} = P \cdot P^{(\ell-1)}$. Define

$$f_{ij} = \sum_{\ell=1}^{4} p_{\ell} P_{\ell i} P_{\ell j} \tag{3}$$

The quantity f_{ij} is interpreted as the probability that a nucleotide is currently in state i in one sequence and state j in the other.

It follows from (2) that

$$f_{ij} = e^{-2\lambda\tau}\{\sum_{\ell=1}^{4} p_\ell [\sum_{k=0}^{\infty} P_{\ell i}^{(k)} \frac{(\lambda\tau)^k}{k!}] [\sum_{k'=0}^{\infty} P_{\ell j}^{(k')} \frac{(\lambda\tau)^{k'}}{k'!}]\} \tag{4}$$

If the matrix P is specified, then the only unknown variable in the expression for f_{ij} is the product $\lambda\tau$. All of the estimates that will be developed are of $\lambda\tau$. In order to estimate λ it is therefore necessary to have an independent estimate of τ.

C. Estimates of $\lambda\tau$ Based on Restriction Map Data

Suppose two homologous DNA sequences are digested with j different restriction enzymes, and that a map is constructed giving the relative locations of the cleavage sites present in each sequence. For simplicity it is assumed that all the recognition sequences are of the same length. The more general case when recognition sequences of different lengths are used will be considered at the end of this section. To avoid repetition the DNA sequences will always be assumed to be linear. The modifications for circular DNA are usually obvious and so they will be omitted.

Let $W = (w_1, w_2, \ldots, w_r)$ be a recognition sequence of length r. The recognition sequence, W, occurs at site i if $Y_i = w_1$, $Y_{i+1} = w_2, \ldots, Y_{i-r+1} = w_r$. Define:

P(W) = probability that W appears at a specific site in a specific sequence

Q(W) = probability that W appears at a specific site in a specific sequence and is not present at that site in the other sequence

R(W) = probability that W appears at the same specific site in both sequences

It follows from the independence of the Y_i within and between sequences that

$$P(W) = \prod_{i=1}^{r} p_{w_i}$$

$$Q(W) = P(W) - R(W)$$

$$R(W) = \prod_{i=1}^{r} f_{w_i w_i}$$

Let

$X_1(W)$ = number of times W is observed at some site in one sequence and not at the same site in the other sequence

$X_2(W)$ = number of times W is observed in both sequences at the same site

$\underline{X}(W) = [X_1(W), X_2(W)]$

It was argued by Kaplan and Langley that $\underline{X}(W)$ has approximately the same distribution as $\underline{Z} = (Z_1, Z_2)$ where Z_1 and Z_2 are independent Poisson variables with means 2MQ(W) and MR(W), respectively. Since Q(W) and R(W) are typically small, M must be large for $\underline{Z}$ to be nonzero with high probability. Furthermore, if J restriction enzymes are used and $\underline{X}(W_1)$, $\underline{X}(W_2)$, ..., $\underline{X}(W_J)$ are the associated vectors of data, then these random vectors can be assumed to be stochastically independent, providing J is not excessive. The likelihood of the data can thus be written as

$$L[\underline{X}(W_1),\ldots,\underline{X}(W_J)] =$$

$$\prod_{j=1}^{J} \frac{[2MQ(W_j)]^{X_1(W_j)} [MR(W_j)]^{X_2(W_j)}}{X_1(W_j)!\, X_2(W_j)!} e^{-M[2Q(W_j) + R(W_j)]}$$

Assuming that the GC content of the sequence is known, Kaplan and Langley (1979) used the matrix in Example 2 and proposed as an estimate of $\lambda\tau$, λ_1, which is that value of $\lambda\tau$ which maximizes L. Although they showed by simulation studies that this estimate performs well, the computational difficulties in finding λ_1 and the need to assume a form for P makes their approach cumbersome.

Kaplan and Risko (1981) exploited the fact that $\lambda\tau$ is typically small and approximated L by replacing Q and R by

$$Q_0(W_j) = P(W_j)(1 - e^{-2\lambda\tau r})$$

and

$$R_0(W_j) = P(W_j)e^{-2\lambda\tau r}$$

The value of $\lambda\tau$ maximizing this approximation of L can be computed explicitly and equals

$$\lambda_2 = -\frac{1}{2r}\log(A)$$

where

$$A = \frac{[(\dot{X}_1 + \dot{X}_2 - B)^2 + 4B\dot{X}_2]^{\frac{1}{2}} - (\dot{X}_1 + \dot{X}_2 - B)}{2B}$$

with

$$\dot{X}_1 = \sum_{j=1}^{J} X_1(W_j) \qquad \dot{X}_2 = \sum_{j=1}^{J} X_2(W_j)$$

$$B = M\sum_{j=1}^{J} P(W_j)$$

When B is estimated by

$$B_0 = \frac{\dot{X}_1 + 2\dot{X}_2}{2}$$ = average number of cleavage sites observed in the two sequences

the estimate λ_2 reduces to

$$\lambda_3 = -\frac{1}{2r}\log\left(\frac{\dot{X}_2}{B_0}\right)$$

The estimate λ_3 has also been derived by Gotoh *et al.* (1979) and Nei and Li (1979) using different arguments. Since the moment method of Nei and Li is short and useful in developing an estimate of $\lambda\tau$ when enzymes coding for recognition sequences of different lengths are used, it is presented here.

Suppose that two classes of enzymes are used, one with recognition sequence of length r(1) and the other of length r(2). Let $\dot{X}_1(1)$ and $\dot{X}_2(1)$ be the associated sums for the first class and $\dot{X}_1(2)$ and $\dot{X}_2(2)$ for the second. Furthermore, let Z(1) and Z(2) denote the number of cleavage sites of lengths r(1) and r(2) present in the ancestral sequence at the time of divergence. Then, to a first approximation,

$$Z(1)e^{-2r(1)\lambda\tau} + Z(2)e^{-2r(2)\lambda\tau} \approx \dot{X}_2(1) + \dot{X}_2(2) \tag{5}$$

The simplest estimate of Z(1) is the average number of cleavage sites of length r(1) in the two sequences, *i.e.*, $B_0(1) = [\dot{X}_1(1) + 2\dot{X}_2(1)]/2$. Similarly, Z(2) can be estimated by $B_0(2) = [\dot{X}_1(2) + 2\dot{X}_2(2)]/2$. Substituting these expressions in (5) results in an equation which can be solved for $\lambda\tau$, and it is this solution which is an estimate of $\lambda\tau$. This estimate is denoted by λ_4. When $\lambda\tau$ is small, the estimate is approximately equal to

$$\hat{\lambda}_4 = \frac{1}{2}\left\{\frac{B_0(1) + B_0(2) - [\dot{X}_2(1) + \dot{X}_2(2)]}{B_0(1)r(1) + B_0(2)r(2)}\right\}$$

This expression for $\hat{\lambda}_4$ has also been given by Engels (1981).

The distribution of λ_2 has been investigated by Kaplan and Risko (1981). They argued from maximum likelihood considerations that λ_2 is approximately normal with mean $\lambda\tau$ and variance $1/I(\lambda\tau)$ where

$$I(\lambda\tau) = (2r)^2 Be^{-2r\lambda\tau}(1 + e^{-2r\lambda\tau})(1 - e^{-2r\lambda\tau})^{-1} \quad (6)$$

It should be noted that this variance is a measure of the variability of λ_2 in a sequence of repetitions of the entire evolutionary process.

When $\lambda\tau$ is small, the variance of λ_2 is approximately $\lambda\tau/4rB$. Consequently, the variance of λ_2 is small when B, the expected number of cleavage sites per sequence, is large. Estimating B by B_0 in (6) results in an expression for the variance which is essentially the same as that given by Nei and Tajima (1981).

Equation (6) easily generalizes to the case when restriction enzymes having recognition sequences of different lengths are used. Indeed, if recognition sequences of two different lengths are used, one obtains

$$I(\lambda\tau) = \sum_{i=1}^{2} [2r(i)]^2 B(i)e^{-2r(i)\lambda\tau}[1 + e^{-2r(i)\lambda\tau}][1 - e^{-2r(i)\lambda\tau}]^{-1}$$

$$\approx [\sum_{i=1}^{2} 4r(i)B(i)]/(\lambda\tau)$$

This approximation of I is denoted by $\hat{I}$.

D. Estimates of $\lambda\tau$ Based on DNA Fragment Data

Although the construction of restriction enzyme maps is straightforward, it is still a laborious task. Consequently, the simpler method of comparing the electrophoretic patterns of the digested DNA sequences has been used to measure genetic diversity (Avise *et al.*, 1979; Brown, 1980).

This kind of data can be summarized by two numbers: S_1, which indicates the number of fragments into which the two DNAs are cleaved; and S_2, the number of fragments conserved. To obtain S_2 it is assumed that comigration implies conservation.

For circular DNA sequences, the number of fragments is the same as the number of cleavage sites, while for linear DNA sequences the number of cleavage sites is one more than the number of fragments. (This assumes that the cleavage sites at the end-points are counted.) If the number of conserved sites is known, then the estimates of $\lambda\tau$ of the previous section can be used. However, it is exactly this information which is sacrificed by not constructing the map. The quantity S_2 is only a lower bound.

Several authors (Upholt, 1977; Nei and Li, 1979; Engels, 1981) have attempted to extract this lost information from S_1 and S_2 by using a mathematical model. Upholt (1977) and Nei and Li (1979) computed the probability that a random fragment is conserved. This approach led to the following equations for the next two estimates of $\lambda\tau$:

$$2S_2/S_1 = \left(\frac{e^{-2\lambda_5 r}}{2 - e^{-\lambda_5 r}} \right)^2 \qquad \text{(Upholt)}$$

and

$$2S_2/S_1 = \frac{e^{-4\lambda_6 r}}{3 - 2e^{-\lambda_6 r}} \qquad \text{(Nei and Li)}$$

In a recent paper Engels (1981) has proposed a nonparametric method for estimating the number of shared sites. For linear DNA, it is not difficult to show that the number of conserved sites equals $S_2 + c$ where c is the number of runs of conserved sites. To estimate c, Engels used a conditional maximum likelihood approach. If it is assumed that all permutations of conserved and unconserved sites are equally likely, then given that S_1 and S_2 are observed, the probability of any value of c can be found by combinatorial considerations. The estimate of c, $\hat{c}$, is then that value which has maximum probability. With an estimate of c in hand, one can estimate the number of conserved sites and compute, say λ_3. The estimate of $\lambda\tau$ obtained in this way will be denoted by λ_7.

The statistical properties of λ_5, λ_6, and λ_7 are unknown.

E. Estimates of $\lambda\tau$ Based on Nucleotide Sequence Data

In a series of papers Kimura and coworkers (Kimura, 1980, 1981; Takahata and Kimura, 1981) have considered the problem of estimating $\lambda\tau$ from nucleotide sequence data. The estimates they develop are all based on moment equations. To illustrate their approach the simplest of their models is considered (Kimura, 1980). Suppose $\lambda = \alpha + 2\beta$ and

$$P = \lambda^{-1} \begin{bmatrix} 0 & \beta & \alpha & \beta \\ \beta & 0 & \beta & \alpha \\ \alpha & \beta & 0 & \beta \\ \beta & \alpha & \beta & 0 \end{bmatrix}$$

where α and β are positive parameters.

Nucleotides at a position that are not in the same state in both sequences can be classified in one of two ways. If the nucleotides in the two sequences are both purines (A,G) or pyrimidines (C,T), then the nucleotide pair is said to be of Type I. The change from one to the other is called a *transition*. All other different nucleotides are of Type II and result from *transversions*. Let P_1 and P_2 denote the probabilities that a pair of different nucleotides at a site is of Type I or of Type II. It follows from standard Markov chain theory (Kimura, 1980) that

$$P_1 = \frac{1}{4} - \frac{1}{2} e^{-4(\alpha+\beta)\tau} + \frac{1}{4} e^{-8\beta\tau}$$

$$P_2 = \frac{1}{2} - \frac{1}{2} e^{-8\beta\tau}$$

The above equations can be solved for $\alpha\tau$ and $\beta\tau$ and consequently $\lambda\tau = \alpha\tau + 2\beta\tau$ can be estimated by

$$\lambda_8 = -\frac{1}{4} \log[(1 - 2\hat{P}_1 - \hat{P}_2)(1 - 2\hat{P}_2)^{\frac{1}{2}}]$$

where $\hat{P}_1$ and $\hat{P}_2$ are the observed frequencies of nucleotide pairs of Type I and Type II. The most general case considered by Takahata and Kimura (1981) allowed for an arbitrary GC content and a more general state dependent $\underline{\lambda}$ vector. For this more complicated model the above type of moment analysis results in equations which in general are too complex to be of use and so simplifying assumptions are necessary. For details the reader is referred to their paper.

An alternative approach to estimating $\lambda\tau$ using nucleotide sequence data has been proposed by Kaplan and Risko (1982). While they make no special assumptions about the matrix P, their derivation assumes that $\lambda\tau$ is small.

Let X_{12} denote the number of nucleotides that are in the same state in both sequences. Then X_{12} has a binomial distribution with parameters M and $f = \sum_{i=1}^{4} f_{ii}$. [See equation (4) for f_{ii}.] The simplest estimate of f is

$$\hat{f} = \frac{X_{12}}{M}$$

This estimate is unbiased, consistent, and since f has an inverse for $\lambda\tau$ small, $\lambda\tau$ can be estimated by λ_9 where

$$\lambda_9 = f^{-1}(\hat{f})$$

An approximate formula for λ_9 can be found in the following way. It follows from the definition of f that

$$\hat{f} \approx e^{-2\lambda_9}[1 + A(\lambda_9)^2] \tag{7}$$

where A is a constant between zero and two depending on the matrix P. If λ_9 is small, $1 + A(\lambda_9)^2 \approx \exp(A\lambda_9^2)$. Substituting this approximation in (7) and solving the resulting equation leads to the equation

$$\lambda_9 = \frac{1 - [1 + A \log (\hat{f})]^{\frac{1}{2}}}{A}$$

Kaplan and Risko (1982) showed that replacing A by 1 introduces negligible error in the estimate, especially if $\hat{f}$ is close to one. To make this substitution it must be true that $1 + \log(\hat{f}) > 0$. If this inequality does not hold, then $\hat{f} > 1/e$, and most likely the approximation in (7) would be too crude.

The statistical properties of λ_9 follow from those of $\hat{f}$. Hence, λ_9 is asymptotically normal with mean $1 - [1 + \log(f)]^{\frac{1}{2}}$ and variance $(1-f)/\{4Mf[1 + \log(f)]\}$.

III. WITHIN POPULATION VARIATION

In this section the inferences that can be drawn from restriction enzyme map data about DNA sequence variation within a population will be investigated. There are two different types of variation that one can estimate from the data. The first is site or nucleotide variation within the sample itself. Estimates of this variation have been proposed by Engels (1981), Ewens *et al.* (1981) and Hudson (1982). The second kind of variation is that within the population. Several reasonable estimates of this variation have been proposed by Engels (1981) and Nei and Tajima (1981). The important problem of relating these measures of population variation to the population parameters suggested by certain population genetics models has also been discussed by Ewens *et al.* (1981) and Nei and Tajima (1981). Many of the ideas of this section are also discussed by Ewens in his chapter of this volume.

For the remainder of this section it is assumed that the restriction enzyme maps have been constructed for n homologous DNA sequences of M nucleotides each. Also, unless otherwise stated. all enzymes used to construct the maps are assumed to code for recognition sequences of length r.

A. Variation in the Sample

Let

ρ = the proportion of sites which are polymorphic in the sample

(Recall that a site is r consecutive nucleotides.) The difficulty with using restriction enzyme map data to estimate ρ is that there is a bias since information is available only about sites which have been cleaved in at least one of the DNA sequences in the sample. Engels (1981), Ewens *et al.* (1981), and Hudson (1982) have proposed the following estimates of ρ which attempt to compensate for this bias:

$$\rho_1 = 1 - \frac{n(m - k)}{c} \qquad \text{(Engels)}$$

$$\rho_2 = \frac{k}{2m} \qquad \text{(Ewens } et\ al.\text{)}$$

$$\rho_3 = \frac{k}{2m - k} \qquad \text{(Hudson)}$$

where

m = number of distinct cleavage sites in the sample

k = number of cleavage sites polymorphic in the sample

c_i = number of members of the sample cleaved at site i, $i = 1,\ldots,m$

$$c = \sum_{i=1}^{m} c_i$$

A discussion of the derivation of these estimates as well as their variances can be found in the chapter by Ewens in this volume.

An estimate of nucleotide polymorphism in the sample, ρ^*, can be obtained from ρ if it is assumed, as suggested by Ewens *et al.* (1981), that any given site can be polymorphic at no more than one of its r positions and all the positions are equally likely. Thus,

$$\rho_i^* = \frac{\rho_i}{r} \qquad i = 1,2,3$$

It is interesting to note that when n = 2, ρ_2^* is approximately equal to $2\lambda_3$. However, it is important to recognize that the interpretation of the two estimates is different. While ρ_2^* is an estimate of the proportion of nucleotides polymorphic in the sample, $2\lambda_3$ is an estimate of the average number of evolutionarily acceptable substitutions per nucleotide and is derived within the context of a specific mathematical model.

B. Population Variation

Several different measures of genetic variation within a population have been suggested by Engels (1981) and Nei and Tajima (1981). Some of these measures and their estimates are given below and the reader is referred to the relevant papers for additional details.

(i) Let p_i denote the frequency of the *ith* restriction enzyme map pattern in the population and x_i its frequency in the sample. Then *restriction map diversity* can be defined as

$$h = 1 - \sum_i p_i^2 \tag{8}$$

This measure corresponds to panmictic heterozygosity in classical genetics. An unbiased estimate of h is

$$\hat{h} = \frac{n}{n - 1}(1 - \sum_i x_i^2)$$

The factor n/(n - 1) arises from the multinomial sampling procedure.

(ii) Let η_{ij} denote the number of site differences between the *ith* and *jth* restriction map pattern. Then

$$\eta = \sum_{i \neq j} p_i p_j \eta_{ij} \tag{9}$$

represents the average number of site differences between two randomly chosen individuals. An unbiased estimate of is

$$\hat{\eta} = \frac{n}{n - 1} \sum_{i \neq j} x_i x_j \eta_{ij}$$

(iii) Let π denote the proportion of different nucleotides between two randomly chosen sequences in the population. Nei and Li (1979) have called π *nucleotide diversity*. Formally, π is defined as

$$\pi = \sum_{i \neq j} p_i p_j \pi_{ij} \qquad (10)$$

where π_{ij} is the proportion of different nucleotides between the i*th* and j*th* restriction map patterns. To estimate π, it is necessary to estimate π_{ij}. If there is little variation between the i*th* and j*th* restriction maps, *i.e.*, $\lambda\tau$ is small, then

$$\pi_{ij} \approx 1 - e^{-2\lambda\tau} \approx 2\lambda\tau$$

Hence any of the estimates of $\lambda\tau$ given in the previous section based on restriction map data or fragment data can be used to estimate π_{ij}. An alternative estimate of π_{ij} based on equation (7) which is more sensitive to larger values of $\lambda\tau$ is

$$\hat{\pi}_{ij} = 1 - e^{-2\hat{\lambda}}(1 + \hat{\lambda})$$

where $\hat{\lambda}$ is an estimate of $\lambda\tau$ based on map or fragment data. An estimate of π is consequently

$$\pi_1 = \frac{n}{n - 1} \sum_{i \neq j} x_i x_j \hat{\pi}_{ij}$$

Another estimate of π has been given by Engels (1981).

His estimate is

$$\pi_2 = \frac{nc - \sum c_i^2}{rc(n - 1)}$$

It is important to remember that π_1 was derived within the context of a specific mathematical model, while π_2 required no specific model assumptions.

As is typically the case, the variances of $\hat{h}$, $\hat{\eta}$, π_1 and π_2 each have two components. The first one is due to sampling and can be obtained from formulae associated with the multinomial distribution. The second one is due to the stochastic changes in the gene frequencies and is dependent on the population model which is used to model the gene frequency process. For additional details the reader is referred to the paper of Nei and Tajima (1981) and the chapter by Ewens in this volume.

C. Estimates of a Population Parameter

An important population parameter which has been suggested by several different genetic theories of evolution, *e.g.*, the infinite alleles model (Kimura and Crow, 1964; Ewens, 1972) and the infinite sites model (Kimura, 1969; Li, 1975; Watterson, 1975) is $\theta = 4N_e\nu$ where N_e is the effective population size and ν is the rate at which existing "alleles" mutate. If the population is haploid instead of diploid, $\theta = 2N_e\nu$.

In the context of the infinite alleles or infinite sites models, the parameter ν must be interpreted, when applied to restriction map data, as the rate at which an existing map mutates to a new one. In this case, the parameter θ is not a natural one since its estimation depends on the restriction enzymes that are used to construct the map. Thus only when the same restriction enzymes are used can the estimates of θ be compared for different populations. A more appropriate parameter would be $\theta^* = 4N_e\lambda$ where λ is the rate of mutation per nucleotide. The connection between θ and θ^* will now be elucidated following Nei and Tajima (1981).

Let W denote a recognition sequence and suppose that it has r_1G's, r_2C's, r_3A's and r_4T's. Then one can write the probability P(W) that W occurs at a site as

$$P(W) = p_1^{r_1} p_2^{r_2} p_3^{r_3} p_4^{r_4}$$

This expression for P(W) assumes that the enzyme recognizes a unique sequence. For those enzymes for which this is not true, P(W) must be modified. This point is well discussed by Nei and Tajima (1981).

To a first order approximation, the rate of loss of W at any site where it is currently present is

$$\sum_{i=1}^{4} r_i \lambda(i) = r\lambda^*$$

where

$$r = \sum_{i=1}^{4} r_i$$

and

$$\lambda^* = [\sum_{i=1}^{4} r_i \lambda(i)]/r$$

Hence, the rate of loss of W in the sequence is

$$MP(W)r\lambda^*$$

At steady state the rate of loss of sites must equal the rate of gain of sites. Thus,

$$\nu \simeq 2MP(W)r\lambda^* \tag{11}$$

If J restriction enzymes are used coding for recognition sequences of lengths r(1), ..., r(J), then the above expression becomes

$$\nu \approx 2M \sum_{j=1}^{J} P(W_j)r(j)\lambda(j)^*$$

$$\approx 2M \sum_{j=1}^{J} P(W_j)r(j)\lambda^{**}$$

where

$$\lambda^{**} = \sum_{j=1}^{J} P(W_j)r(j)\lambda(j)^* / \sum_{j=1}^{J} P(W_j)r(j)$$

For the remainder of this section it will be assumed for simplicity that all the $\lambda(i)$ are equal and so

$$\theta = \theta^*[2M \sum_{j=1}^{J} P(W_j)r(j)]$$

The multiplier $2M \sum_{j=1}^{J} P(W_j)r(j)$ can be estimated by $2 \sum_j d(j)r(j)$ where $d(j)$ denotes the number of times W_j appears on average in the sample. Hence, any estimate of θ can be used to estimate θ^*.

If only a few restriction map patterns are observed in the sample, and if it is assumed that all patterns are selectively neutral, then one is tempted to estimate θ within the context of the infinite alleles model. Indeed it follows from Ewens (1972) that an estimate of θ, θ_1, is the solution of the equation

$$k_0 = \sum_{j=0}^{n-1} \frac{\theta_1}{j + \theta_1}$$

where k_0 denotes the number of distinct map patterns observed in the sample.

Alternative estimates of θ can be obtained from the estimates of population genetic variability of the previous section. The next three estimates are given by Nei and Tajima (1981).

(i) If there is no selection, the mean and variance of h [equation (8)] are [Nei and Tajima (1981)]

$$E(h) = \frac{\theta}{1 + \theta}$$

and

$$\sigma^2(h) = \frac{2\theta}{(1 + \theta)^2(2 + \theta)(3 + \theta)}$$

Hence, another estimate of θ is

$$\theta_2 = \frac{\hat{h}}{1 - \hat{h}}$$

(ii) If an infinite sites model is assumed to describe the gene frequency process, then, from equation (9)

$$E(\eta) = \theta$$

and

$$\theta^2(\eta) = \theta + \theta^2$$

Thus an estimate of θ is

$$\theta_3 = \hat{\eta}$$

(iii) Still another estimate of θ can be obtained from π [equation (10)] by noting that for the infinite-sites model

$$E(\pi) = \theta^*$$

Thus,

$$\theta_4 = 2[\sum_{j=1}^{J} d(j)r(j)]\hat{\pi}$$

where $\hat{\pi}$ is an estimate of π.

An alternative estimate of θ^* has been proposed by Ewens *et al.* (1981). Assuming that a Wright-Fisher model with mutation and no selection describes the frequency process of a particular nucleotide, they showed by a diffusion approximation that, for the case when all recognition sequences are of the same length,

$$\theta^* = \frac{\rho}{r \log (n)}$$

Consequently, their estimate of θ is

$$\theta_5 = \frac{2[\sum d(j)]\hat{\rho}}{\log (n)}$$

where $\hat{\rho}$ is an estimate of ρ.

IV. SIMULATION AND DATA ANALYSIS

A. Simulations

The best way to compare the various estimates of $\lambda\tau$ that have been proposed in the earlier part of this paper is to use simulation methods. Since each nucleotide is assumed to evolve independently within and between sequences, it is sufficient, for simulating the two homologous sequences, to simulate M independent random variables each with distribution given by the f_{ij} in equation (4).

For the simulations in Tables 2-4, M is set at 16000, which is the approximate length of mtDNA. The eleven restriction enzymes used to construct the maps and their recognition sequences are given in Table 1. The three different choices for P given in Examples 1-3 were used to determine the f_{ij}. For each set of conditions one hundred simulations were performed and the averaged estimates are given in the tables.

The results of the simulations are given in Tables 2-4. The estimate λ_1 was not included in the tables since its computation is too difficult for practical use. All of the estimates perform well for small $\lambda\tau$ regardless of the choice of P. The robustness of the estimates to different choices of P is explained by the observation that for small $\lambda\tau$, few nucleotides mutate and so the effects of the different stochastic structures are negligible. In Table 3, a similar pattern is obtained if the GC content is set equal to 0.3 or 0.7, while in Table 4 a similar pattern is obtained if the other two sets of parameter values discussed by Takahata and Kimura (1981) are used.

Table 1. The eleven restriction enzymes used for the simulation study and their recognition sequences.

	Enzymes	Recognition Sequence
1.	*Eco* RI	GAATTC
2.	*Bam* HI	GGATCC
3.	*Hind* III	AAGCTT
4.	*Xba* I	TCTAGA
5.	*Pst* I	CTGCAG
6.	*Hpa*	GTTAAC
7.	*Kpn* I	GGTACC
8.	*Bgl* II	AGATCT
9.	*Pvn* II	CAGCTG
10.	*Xho* II	CTCGAG
11.	*Sst* I	GAGCTC

The effect of different choices for P begins to show as $\lambda\tau$ increases. The estimates based on sequence data, λ_8 and λ_9, appear to perform best. This is not surprising since nucleotide sequence data contain the most information about the sequences. It should be noted that the excessive length of the sequence under study does not appear to influence the quality of the estimates λ_8 and λ_9. Kaplan and Risko (1982) arrived at similar results using sequences of only 300 nucleotides.

Estimates of $\lambda\tau$ based on restriction enzyme map data performed somewhat poorer than those based on sequence data while estimates based on fragment data did the worst. It is interesting to note that $\lambda\tau$ must be small, say less than 0.1, before the fragment-based estimates are reliable.

Table 2. Simulation study of the various estimates (and sample standard deviation) of $\lambda\tau$. The transition matrix in Example 1 was used.

	True Value of $\lambda\tau$				
	.01	.05	.10	.20	.30
λ_2	.0100 (.0003)	.0484 (.0010)	.0995 (.0018)	.1918 (.0043)	.2593 (.0047)
λ_3	.0100 (.0003)	.0477 (.0010)	.0989 (.0018)	.1913 (.0041)	.2584 (.0048)
λ_5	.0098 (.0002)	.0048 (.0013)	.0968 (.0023)	.1220 (.0028)	.1349 (.0012)
λ_6	.0097 (.0003)	.0466 (.0012)	.0928 (.0021)	.1168 (.0026)	.1290 (.0013)
λ_7	.0095 (.0003)	.0473 (.0012)	.0954 (.0022)	.1209 (.0027)	.1337 (.0019)
λ_8	.0099 (.0006)	.0501 (.0015)	.0995 (.0020)	.1988 (.0029)	.2974 (.0044)
λ_9	.0099 (.0005)	.0505 (.0015)	.1010 (.0021)	.2052 (.0032)	.3123 (.0048)

Table 3. Simulation study of the various estimates (and sample standard deviation) of $\lambda\tau$. The transition matrix in Example 2 was used, with GC content equal to 0.5.

	True Value of $\lambda\tau$				
	.01	.05	.10	.20	.30
λ_2	.0099 (.0003)	.0508 (.0009)	.0955 (.0016)	.1835 (.0034)	.2575 (.0054)
λ_3	.0099 (.0003)	.0507 (.0010)	.0960 (.0017)	.1823 (.0034)	.2573 (.0053)
λ_5	.0099 (.0004)	.0513 (.0012)	.0917 (.0023)	.1291 (.0020)	.1356 (.0010)
λ_6	.0098 (.0004)	.0497 (.0011)	.0880 (.0021)	.1235 (.0018)	.1296 (.0010)
λ_7	.0097 (.0004)	.0508 (.0012)	.0901 (.0022)	.1277 (.0020)	.1327 (.0012)
λ_8	.0100 (.0006)	.0499 (.0015)	.0999 (.0021)	.2001 (.0033)	.2989 (.0040)
λ_9	.0100 (.0006)	.0503 (.0015)	.1016 (.0022)	.2070 (.0035)	.3152 (.0045)

Table 4. Simulation study of the various estimates (and sample standard deviation) of $\lambda\tau$. The transition matrix in Example 3 was used, with α = 0.00022, β = 1.36α, γ = 2.19α, ε = 1.02β, δ = 1.02α.

	True Value of $\lambda\tau$				
	.01	.05	.10	.20	.30
λ_2	.0111 (.0004)	.0559 (.0011)	.1085 (.0018)	.2113 (.0042)	.2817 (.0042)
λ_3	.0113 (.0004)	.0567 (.0013)	.1081 (.0018)	.2120 (.0042)	.2826 (.0040)
λ_5	.0166 (.0004)	.0573 (.0015)	.1014 (.0022)	.1305 (.0023)	.1295 (.0042)
λ_6	.0115 (.0004)	.0555 (.0015)	.0972 (.0021)	.1248 (.0022)	.1239 (.0040)
λ_7	.0114 (.0004)	.0566 (.0015)	.1001 (.0022)	.1288 (.0024)	.1313 (.0037)
λ_8	.0114 (.0006)	.0567 (.0016)	.1179 (.0021)	.2249 (.0032)	.3377 (.0050)
λ_9	.0114 (.0006)	.0571 (.0016)	.1149 (.0022)	.2330 (.0034)	.3572 (.0056

The behavior of the estimate of the variance of λ_3 has been studied via simulation by Kaplan and Risko (1981). They showed that for $\lambda\tau$ small, the estimate performs quite adequately.

In Table 5 simulation results are presented on the behavior of the estimate λ_4. For this simulation four enzymes having recognition sequences six nucleotides long and two enzymes having recognition sequences four nucleotides long were used. The results in Table 5 indicate that this estimate and the estimate of its standard deviation both are adequate for small $\lambda\tau$.

B. Data Analysis

Langley, Montgomery and Quattlebaum (1982) constructed the restriction maps of a DNA sequence 12,000 base pairs long containing the Adh locus of *Drosophila melanogaster* for eighteen wild caught *D. melanogaster* and five closely related species. Eight enzymes were used in the survey, all having a unique hexanucleotide recognition sequence. For the purposes of this chapter only the substitutional differences between the maps will be considered and the insertion-deletion variation ignored.

Table 5. Simulation study of λ_4 (and sample standard deviation) and $I^{-1}(\lambda_4)$. For the simulation the transition matrix in Example 2 was used with GC content equal to 0.5. The restriction map was constructed using two enzymes coding for sequences of length 4 and four of length 6. The coding sequences were CCGG, GGCC, AAGCTT, GAATTC, TGTACA, and CTGCAG.

	True Value of $\lambda\tau$				
	.01	.05	.10	.20	.30
λ_4	.0097 (.0021)	.0492 (.0069)	.0964 (.0099)	.1860 (.0212)	.2709 (.0281)
λ_4	.0093 (.0019)	.0400 (.0045)	.0654 (.0044)	.0927 (.0042)	.1055 (.0030)
$[I^{-1}(\lambda_4)]^{\frac{1}{2}}$	.0020	.0053	.0087	.0166	.0262
$[I^{-1}(\lambda_4)]^{\frac{1}{2}}$	.0019	.0041	.0052	.0062	.0066

For the eighteen wild caught *D. melanogaster*, twenty monomorphic cleavage sites and four polymorphic ones were detected. The numbers of chromosomes in the sample cut at each of the polymorphic sites are 11, 3, 3, and 12, respectively. It should be noted that for one of the sequences in the sample it could not be determined if it had a *Hind* III site at positions -3 or at -2.7. For definiteness it is assumed that the site occurred at position -3. The three estimates of ρ and ρ^* are given in Table 6. It is interesting to note that all the estimates are comparable.

Six distinct restriction map patterns were detected in the sample and the frequencies of these patterns are given in Table 7. The estimate of restriction map diversity is 0.816. To compute the mean number of restriction site differences, it is convenient to make a table of the η_{ij}. These are given above the diagonal of Table 8. From the data in Table 8, $\hat{\eta}$ was computed to be 1.6.

Below the diagonal Table 8 contains the estimates of the π_{ij}. These values were estimated by $2\lambda_3$. Based on these numbers, π_1, an estimate of nucleotide diversity, was found to be .006. The other estimate of nucleotide diversity, π_2, also equals .006.

Table 6. Estimates of ρ and ρ^*.

ρ_1 = .1	ρ_1^* = .016
ρ_2 = .074	ρ_2^* = .012
ρ_3 = .11	ρ_3^* = .018

Table 7. Frequency of the six restriction map patterns.

1	2	3	4	5	6
$\frac{4}{18}$	$\frac{4}{18}$	$\frac{2}{18}$	$\frac{1}{18}$	$\frac{3}{18}$	$\frac{4}{18}$

Table 8. Estimates of the η_{ij} and π_{ij}[a].

	1	2	3	4	5	6
1	0	1	2	1	2	1
2	.004	0	1	2	3	2
3	.008	.004	0	1	4	3
4	.004	.008	.004	0	3	2
5	.008	.012	.016	.01	0	1
6	.004	.008	.012	.016	.004	0

[a]The upper diagonal values are the $\hat{\eta}_{ij}$ and the lower ones are the $\hat{\pi}_{ij}$.

The various estimates of θ and θ^* are given in Table 9. All of the estimates of θ, except θ_2, are comparable. The behavior of θ_2 is not particularly surprising since it is known for the single locus case that heterozygosity has very poor sampling properties and in general tends to overestimate θ (Zouros, 1979).

Table 9. Estimates of θ and θ^* [a].

$\theta_1 = 2$	$\theta_1^* = .008$
$\theta_2 = 6$	$\theta_2^* = .02$
$\theta_3 = 1.6$	$\theta_3^* = .006$
$\theta_4 = 1.55$	$\theta_4^* = .006$
$\theta_5^b =$ 1.08, 1.62, 1.48	$\theta_5^* =$ .004, .006, .006

[a] The estimate of $2M \sum_{j=1}^{J} P(W_j)r(j)$ is 259.

[b] The three values of θ_5 correspond to the three estimates of ρ in Table 6.

V. CONCLUSIONS

The advent of recombinant DNA technology has resulted in an ever-growing body of data describing genetic differences between organisms at the fundamental level: the nucleotide sequence. The need to summarize this large body of data in terms of easily computed, meaningful quantities has prompted the development of new statistical techniques.

The first part of this chapter discusses the problem of estimating K, the average number of evolutionarily acceptable substitutions per nucleotide between two homologous DNA sequences which have diverged from a common ancestor. Estimates of K are reviewed which are based on restriction map data, restriction fragment data, and nucleotide sequence data. All the estimates are derived within the context of a specific mathematical model which describes the evolutionary process at the nucleotide level. In order to make the analysis tractable, several simplifying assumptions are made in constructing the model.

One basic assumption is that nucleotides within a sequence evolve independently of each other. While this assumption is surely not globally correct, the derived estimates will still be adequate so long as the departure from independence is not severe. It is difficult to study this issue by simulation since it is not clear how to simulate sequences where nucleotides do not evolve independently.

It is also assumed that sequence variation is caused only by base pair substitution and not by insertions or deletions. In many cases this assumption is inappropriate since much of the variation is, in fact, due to small insertions and deletions (Shen *et al.*, 1981). In the extreme case insertions and deletions may make it very difficult to identify conserved cleavage sites, and, consequently, the estimates of K will be biased.

Another assumption is that the rates of substitution are constant in time. When the substitution rates are small, this assumption is a reasonable one to make since temporal fluctuations in the substitution rates are probably small and there is no reason to believe they are biased.

Rates of substitution are also assumed to be the same for all nucleotide positions in the sequence. This assumption does not hold in general since there is evidence that mutation rates vary from gene to gene in a genome and even from region to region within a gene (Nei, 1975). Certainly, within a coding region, it is more appropriate to estimate rates of substitution for first, second and third position nucleotides separately. If the number of substitutions per nucleotide is small, however, there is most likely little error introduced in assuming equal mutation rates.

All of the estimates of K with the exception of K_1 $(2\lambda_1)$ require no assumptions about the matrix P other than that it satisfies equation (1). The reason for this is that all the estimates are derived under the assumption that $\lambda\tau$ is small. Consequently, if the observed variation between the two sequences is large, the resulting estimates of K are probably biased. An obvious way to compensate for this bias is to assume an explicit form for the matrix P and derive a more complicated estimate of K, *e.g.*, K_1. The problem with this approach is that there is in general no empirical information about P.

The second part of this chapter studies estimates of genetic variation within a sample and within a population which are based on restriction map data. While the estimates of K are developed from a specific substitution model, the estimates of variation require no such structure.

Implicit in the derivations of all the estimates based on restriction map data and fragment data is the assumption that cleavage sites can be accurately identified. In practice this is not always the case because of experimental difficulties. For example, two unique sites on different chromosomes which are at nearby locations on the chromosomes could easily be identified as the same site. Thus, there is probably a tendency to underestimate K. Estimates based on fragment data are particularly sensitive to this identification problem, especially if there are many small fragments. Aquadro and Greenberg (1983) give an example of just how bad estimates based on fragment data can be.

Despite the potential defects of the methods described in this chapter, they are very useful in that they provide a simple and concise way of analyzing and summarizing this rapidly growing body of data.

Chapter 5

ANALYSIS OF VARIATION IN RELATED DNA SEQUENCES

A. H. D. Brown
*Division of Plant Industry, CSIRO,
Canberra City,
Australian Capital Territory, Australia*

and

Michael T. Clegg
*Departments of Molecular and Population Genetics, and Botany,
University of Georgia,
Athens, Georgia*

I. INTRODUCTION

Recombinant DNA technology is dominating biological research in the eighties. A fundamental advance afforded by this technology is the ability to sequence defined fragments of DNA. It is then feasible to compare sequences of two homologous DNA fragments within and between individuals or species. The comparison of complete DNA sequence data has provided, and promises to continue to provide, an unprecedented view of the elemental processes of evolutionary change. To date, such sequence comparisons have been applied to duplicate genes of known functions (*e.g.*, cytochrome C in yeast, Montgomery *et al.*, 1980; globin sequences in mice, Konkel *et al.*, 1980). In these studies the nucleotide sequence for each member of a duplicate pair of genes is compared to identify regions of the molecule which are conserved, versus regions which are permissive of evolutionary change. As the application of sequencing technology expands, numerous copies of a particular sequence will be available for comparison. For instance, different copies of a gene which is highly reiterated in the genome may be sequenced, or several copies of a particular single copy sequence drawn from different individuals may be analyzed. In this chapter we will consider the analysis of sequence variation in a sample of highly repeated genes. Our objective will be to illustrate several statistical questions which arise naturally from such data.

II. THE DATA: SATELLITE SEQUENCE VARIATION IN ZEA MAYS

A. Source and Definitions

Knob heterochromatin in maize has, as a major constituent, a particular DNA sequence of about 180 base pairs (Peacock *et al.*, 1981; Dennis and Peacock, 1982). This DNA sequence is very highly repeated with up to 10^6 tandem copies in the largest knob on chromosome 10. Thus, there are over 10^6 copies of this sequence in the maize genome. Independent copies of the sequence were drawn from a cloned library of the maize genome. Because the number of copies is very large, the probability of drawing two clones of the same copy can be regarded as effectively zero.

Each copy has the permutation GGCC once. This unique sequence is the recognition site for the restriction endonuclease *Hae* III which cleaves the DNA molecule whenever GGCC occurs. The *Hae* III site was the cloning site, and we will arbitrarily regard this site as the starting point for each repeating unit.

We have used the terms *sequence* and *copy*. While the meaning of these terms is usually clear from the context, we will define them more formally. The term *sequence* refers to a general class of genes which have essentially the same succession of nucleotides or nucleotide sequence or DNA sequence; that is, the particular permutation or linear order of nucleotides in a DNA fragment. Authors have used the word "sequence" either to denote such families collectively or to refer to one particular example (repeat or copy). We use the term *copy* to refer to a single DNA sequence or a single sample realization. In the present context, we will refer to the nucleotide sequence obtained from a single 180 base pair (bp) repeat as a copy.

Returning to the data at hand, 11 clones were drawn and the complete nucleotide sequence was determined for each clone. Of these 11 clones, two occurred as 360 bp fragments because of mutations in the *Hae* III recognition sequence. One of the 360 bp clones was completely sequenced (yielding two copies) and one was half sequenced (yielding one copy). We therefore have 12 copies of a 180 bp sequence. Because the basis for selecting fragments was the *Hae* III recognition site, we exclude this site from consideration in our analysis of DNA sequence variation, thereby reducing each copy to 176 bp.

For many purposes it is convenient to take one sequence as a standard of reference. For example, we may wish to consider the number of mutational substitutions among copies relative to a reference nucleotide sequence. The choice of a reference is arbitrary. It may be the most common nucleotide sequence among the copies, or the ancestral sequence from which all copies are derived. We have taken as a reference an arbitrary nucleotide sequence constructed by placing at the i*th* position ($i = 1, 2, \ldots, 176$) the most frequent nucleotide at that position over all 12 copies, as has been done by other workers (*e.g.*, Miklos and Gill, 1981). We call this the *consensus* (nucleotide) sequence. Two of the 12 copies were observed to have exactly the consensus sequence. Figure 1 shows the distribution of nucleotide substitutions relative to the consensus sequence among all 12 copies. Our task is to analyze the

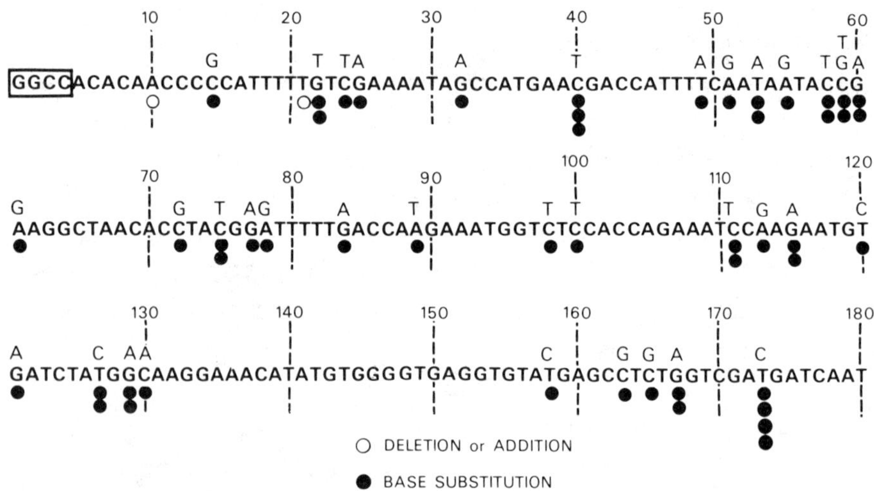

Figure 1. The consensus nucleotide sequence of maize satellite DNA associated with knob heterochromatin. Dots below the sequence indicate the position and frequency of variants, and the variant nucleotide at the site is indicated above the sequence.

patterns of nucleotide substitutions and relationships shown in Figure 1.

B. Kinds of Variation in DNA Sequences

To proceed with an analysis of the pattern of nucleotide substitutions we must first classify the various mutational events which can be distinguished. Table 1 lists the kinds of variants in comparison with the consensus sequence. Considering variants at a single base, if the consensus sequence is known then 12 types of base substitutions are defined. Of these, four are transitions (one purine is replaced by the other, or one pryimidine is replaced by the other). The remaining eight are transversions (where a purine is replaced by a pyrimidine or vice versa). Transversions can be further classified into those which preserve the number of hydrogen bonds (*i.e.*, replace a base by its pairing complement), and those which alter the number of hydrogen bonds (as do all transitions) by either increasing (A → C; G → T) or decreasing them.

Table 1. Kinds of sequence variation.

<u>At Single Sites</u>		
(1) Transitions		A ⇄ G, C ⇄ T*
(2) Transversions	(a) preserve hydrogen bond	A ⇄ T, G ⇄ C
	(b) alter hydrogen bonding	A ⇄ C, G ⇄ T
(3) Additions		
(4) Deletions		
<u>At Several Contiguous Sites</u>		
(1) Additions		
(2) Deletions		
which may be classified, first, on the number of sites involved:		
	(a) one or more triplets of nucleotides	
	(b) other than triplets	
and, second, on the nucleotide sequence:		
	(a) tandem repeat	
	(b) reverse repeat	
	(c) unrelated to nearby flanking sequences, but occurring elsewhere in the genome	
	(d) unique sequence	
	(e) complex combination of the above	

*Where Adenine (A) and Guanine (G) are purines;
Cytosine (C) and Thymine (T) are pyrimidines.
Adenine forms two hydrogen bonds to pair with thymine.
Guanine forms three hydrogen bonds to pair with cytosine.

In translated regions, the magnitude of the effects of additions or deletions will depend on how early in the sequence of the gene the change occurs (*i.e.*, how close to the 5' end of the gene), and whether there is downstream from the variant a compensating change which restores the reading frame. Hence, it may be useful to classify such variants as to where in the sequence they occur.

In general then, the modes of variation in sequence include the number of bases, the types of bases involved and the position within the gene or larger segment of DNA. All these events are distinguishable from a knowledge of the nucleotide sequences. This situation stands in marked contrast to electrophoretically detectable variation of proteins, and less so, to gain or loss of restriction sites. Several kinds of events are resolvable at the DNA sequence level but may not be distinct at other levels.

For this reason, the amount of information and the potential to uncover new phenomena of variation are greatly increased. In view of the phenomenological wealth of this new kind of data, it is likely that several statistical procedures will be needed for a complete analysis. No single summary statistic such as the observed number of alleles at a locus (see chapter by Ewens in this volume) or the probability of identity for two copies of a gene is likely to be adequate in extracting all information from the data. Our view is that several approaches should be taken to the analysis.

III. KINDS OF BASE SUBSTITUTIONS

Each copy was compared with the consensus sequence and all single base mutations were classified. Table 2 summarizes the results. At each site, each distinctly different change was counted. Only one addition and two deletions were found which were too few for any trend in this class to be evident. An analysis of trends in base substitution, however, is possible.

Table 2. Observed single base differences from the consensus maize satellite sequence.

Consensus Sequence		Base Introduced				
Base Replaced	Number	A	C	G	T	Total
A	58	-	0	5	1	6
C	38	1	-	5	8	14*
G	37	9	0	-	1	10
T	43	2	4	0	-	6
Totals	176	12	4	10	10	36

*C substituted twice at position 59.

The first test is for any bias in the bases replaced. The expected frequencies were computed from the frequency of bases in the consensus sequence and are compared with the observed numbers in Table 3. The chi-square test of whether the observed frequencies agree with these expectations indicated a significant departure. When the three possible pairwise groupings of the bases are made, a very striking heterogeneity in base replacement is detected. Either member of the base pair (cytosine, guanine) which forms three hydrogen bonds is significantly more likely to be the site of base substitutions than the 2H-bond pair (adenine and thymine). The (C, G) pair is about 2.7 times more likely to be replaced.

Table 3. Observed and expected numbers of bases replaced from consensus sequence.

Base Replaced	Observed Number	Expected Number
Adenine	6	11.86
Cytosine	14	7.77
Guanine	10	7.57
Thymine	6	8.80

Component	d.f.	Chi-square
Purines vs Pyrimidines	1	1.32
2H-pairs vs 3H-pair	1	8.52 ($P < 0.005$)
A + C vs G + T	1	0.02
Total	3	9.56 ($P < 0.025$)

Next we can ask whether the new mutant bases have been accepted in accord with random expectation. If the expectations are made conditional on the original base composition, the expected number of A's in all bases introduced is $36(38 + 37 + 43)/(176 \times 3) = 8.05$, assuming events are equally likely. Similarly, the other expectations were computed and the resulting chi-square analysis is shown in the columns denoted by (i) in Table 4. There is some bias towards introducing the 2H-bond pairs more frequently (about 1.7 times), but this was not statistically significant.

However, the first analysis already established that there was a substantial bias in the kind of base replaced. It could be argued then that the test of bases introduced should use expectations conditional on the observed frequency of bases replaced. The expected number of A's would then be (14 + 10 + 6)/3 = 10. The other expected numbers and the chi-square analysis are given in the columns headed (ii) in Table 4. These results are much closer to the observed frequencies than in the first conditioning (i). Hence the observed bias in the bases introduced is readily explained as a consequence of the observed bias in the sites of substitution.

Table 4. Observed and expected numbers of bases introduced.

Base Introduced	Observed Number	Expected Number* (i)	Expected Number* (ii)
Adenine	12	8.05	10.00
Cytosine	4	9.41	7.33
Guanine	10	9.47	8.67
Thymine	10	9.07	10.00

Component	d.f.	Chi-square (i)	Chi-square (ii)
Purines vs Pyrimidines	1	2.23	1.23
2H-pair vs 3H-pair	1	2.65	0.45
A + c vs G + T	1	0.24	0.20
Total	3	5.15	2.12

*Expectations conditioned on
(i) observed frequency of bases in consensus sequence and
(ii) observed frequency of bases replaced.

Finally, we turn to an analysis of the frequency of kinds of substitutions. For this we regard Table 2 as a 4 x 4 incomplete contingency table (Bishop *et al.*, 1975), incomplete because replacements of each base with itself are undetectable. The null hypothesis for the contingency chi-square test is that the expected frequency for each of the 12 particular kinds of substitution is obtained simply as the product of the marginal frequency of base replacement and a constant probability for each kind of substituting base. Unfortunately, neither of these quantities is directly observed as a marginal total because of the four missing cells.

Let us denote the four cells as

$$f_A = \text{number of times } A \to A$$

with similar definitions for f_C, f_G and f_T. Further, let

$$(p_i: i = A, C, G, T;\ 0 \leq p_i \leq 1;\ \Sigma p_i = 1)$$

denote the probability that the newly introduced base is A, C, G or T, respectively, and k denote the total number of sites of substitution including both the observed (36) and unobservable changes. If

$$(n_i;\ i = A, C, G, T)$$

are the numbers of *observed* frequencies of newly introduced bases (in this example, $n_A = 12$, etc., and $\Sigma n_i = 36$), then

$$f_i = k\, p_i - n_i$$

If

$$(m_i;\ i = A, C, G, T)$$

are the *observed* numbers of sites of substitution (in this example $m_A = 6$, etc., $\Sigma m_i = 36$), then p_i can be expressed as the solution of the quadratic

$$k\, p_i^2 - (k + n_i - m_i)\, p_i + n_i = 0$$

$$p_i = \{(k + n_i - m_i) \pm [(k + n_i - m_i)^2 - 4\, kn_i]^{\frac{1}{2}}\} / 2k$$

The iterative solution for k which satisfies the constraints is

$$k = 47.22$$

for which the p's and f's are

$$p_A = 0.3118, \; p_C = 0.1284, \; p_G = 0.3045, \; p_T = 0.2553$$

$$f_A = 2.7232, \; f_C = 2.0630, \; f_G = 4.3785, \; f_T = 2.0505$$

From these values, the expected frequency for the ij*th* cell in Table 5 follows as

$$(m_i + f_j)\, p_j$$

The chi-square value for the comparison of the observed frequencies (Table 2) with these expected frequencies is

$$\chi^2 = 30.89 \quad \text{(5 degrees of freedom, } P < 0.005\text{)}$$

Therefore, the hypothesis that each base is substituted with a particular probability independent of the kind of base it is replacing must be rejected in this case.

Table 5. Expected numbers of substitutions, after taking account of marginal frequencies and unobservable classes.

Base Replaced	Base Introduced				Total
	A	C	G	T	
A	-	1.12	2.65	2.23	6
C	5.01	-	4.89	4.10	14
G	4.48	1.85	-	3.67	10
T	2.51	1.03	2.46	-	6
Total	12	4	10	10	36

To analyze the source of this bias further, Table 6 reduces the comparisons first by combining complementary substitutions, *i.e.*, a change of A → G would entail T → C in the opposite stand. Presumably, the change T → C in the reference strand is subject to similar constraints as the change A → G. The chi-square value is now 25.76 (P < 0.005), still significantly high. The six classes can be grouped into types of substitutions as defined in Table 1. The chi-square test, with two degrees of freedom, shows that each type has a distinctive occurrence. Transitions are about five times as likely as are transversions. Transversions which alter the number of hydrogen bonds are at a relative disadvantage of 1/8 to those which preserve the number.

Table 6. Analysis of base substitutions.

Substitution	Type		Observed	Expected
A → G, T → C	Transitions		9	3.7
C → T, G → A	Transitions		17	8.6
A → T, T → A	Transversions	ΔH = 0	3	4.7
C → G, G → C	Transversions	ΔH = 0	5	6.7
A → C, T → G	Transversions	ΔH ≠ 0	0	3.6
C → A, G → T	Transversions	ΔH ≠ 0	2	8.7

	d.f.	Chi-square	
Between six classes	4	25.76	(P < 0.005)
Three types of substitution	2	25.04	(P < 0.005)

The overall conclusion from this section is that base substitution is not random as to kind of replacement. Cytosine and guanine are the preferred sites of substitution (2.5 times), and given that bias, transitions are the preferred kind of substitutions (5 times). The net effect is an overall decrease in the capability of H bond formation. It is interesting that Vogel and Kopun (1977) found a higher incidence of transitions as compared with their expectation if the direction of mutation were random for translated portions of the genome. They also found that substitutions affecting the G, C pair were significantly more frequent than those affecting the A, T pair.

IV. CLUSTERING OF SITES OF CHANGE ALONG A SEQUENCE

We now turn to the distribution of the sites of changes within the linear sequence. The general question under study is whether mutable sites are randomly distributed along the segment of DNA, or whether there are regions which are relatively conserved and others where the sites of variation are clustered. In many ways this question resembles the study of runs of various kinds of species along a transect in plant ecology (Pielou, 1977).

For the case of knob heterochromatin sequences in *Zea mays*, the map of *mutable* sites is shown in Figure 1. Sites at which no variants from the consensus sequence were observed in the sample of cloned fragments are designated here as *constant*. The starting point can be taken at any position in the sequence because the sequence is tandemly repeated many times in the genome.

A first approach to the pattern of mutable sites is to divide the sequence into sectors of consecutive "constant" sites terminated by a mutable site and form the observed distribution. Because of the arbitrary starting point, positions 174 through 14 form one such sector. The random variable is the number, or run length (r), of adjacent constant sites before the next mutable site is reached. This distribution is given in Table 7. If occurrence of a mutable site is random and equivalent to that of a success in a Bernoulli trial, the distribution should be geometric in character. The expected value of each class is npq^r, where n (=35) is the total number of mutable sites, p is the constant probability that a site is mutable and q = 1 - p. The maximum likelihood estimate of p is 35/176 = 0.1989, and using this estimate, the expected distribution is given in Table 7 where consecutive classes are grouped to give expected values exceeding unity. Despite a marked excess in the second class, the chi-square value for the goodness of fit was 12.8, which for 11 degrees of freedom indicates an acceptable fit ($0.25 < P < 0.5$). Overall, the observed distribution agrees with the null hypothesis of random location of mutable sites. However, the variance of the observed distribution (30.03) significantly exceeds that of the fitted geometric distribution ($q\,p^{-2} = 20.26$; $\chi^2_{(34)} = 50.4$; $0.01 < P < 0.05$). A single conserved sector of unusual length (positions 131 to 157) is evident. Further, the apparent clustering of mutable sites in the region 51-61 is suggestive.

Table 7. Distribution of runs of invariant sites.

Run Length	Observed Frequency	Expected Geometric Frequency
0	7	6.96
1	11	5.58
2	2	4.47
3	0	3.58
4	3	2.87
5	3	2.30
6	1	1.84
7	2	1.47
8	2	1.18
9	0	1.70 (9 and 10 combined)
10	2	
11,12	0	1.09
13 - 16	1	1.15
over 16	1	0.81

A more formal approach to these questions is afforded by renewal theory of recurrent processes. To answer questions of this first kind, Feller (1968, p. 324) shows that the mean and variance of the recurrence times of runs of length r are (in our notation)

$$\mu = p^{-1}\,q^{-r} - p^{-1}$$

$$\sigma^2 = (pq^r)^{-2} - (1 + 2r)\,p^{-1}\,q^{-r} - qp^{-2}$$

If a long sequence (n bases) is examined, then for large n, the number N of runs of length r produced is approximately normally distributed. The approximate 95 percent limits for N are

$$n\mu^{-1} \pm 2\sigma(n\mu^{-3})^{\frac{1}{2}}$$

In the case of our example (r = 27, p = 0.1989), these confidence limits are (0, 0.68) so that this run is a statistically significant departure (at least at the 95 percent level) from random expectation. Interestingly, Dennis and Peacock (1982) have sequenced clones from homologous DNA from the related species *Zea mexicana* and so far no variant has been encountered in the region.

If this run is now removed from consideration, a sequence of 148 nucleotides remains. The question now is whether there is any further departure from randomness. In particular, is there a "hot spot" of mutations in the region 51-61? The same approach is applicable and can be extended to formulating the recurrence time for the following more general situations.

The question we want to ask is what is the expected total run length (waiting time) for a sequence of at least k-1 or at least k-2 unordered mutable sites in an interval of k nucleotides? First, two kinds of "successful" events (each consisting of k nucleotides) can be defined: a completely "closed" k-tuple is one where the successful event both begins and ends with a mutable site, and a semi "open" k-tuple is one where the successful event begins with a mutable site but may or may not end with a mutable site.

In order to apply the theory of recurrent events, it is necessary to define the occurrence of successive "successful" events as nonoverlapping. Thus the first occurrence of the event of length k is uniquely defined, and we now agree to start counting from scratch every time the full event occurs.

In the case of the successful event including either k or k - 1 mutable sites, the mean recurrence time for the "closed" case is

$$\mu[k;\ k\text{-}1,\ k] = \frac{1}{p} + \frac{1 + (k-3)\,q + \sum_{i=0}^{k-3} p^{i}[p^{2} + i^{2}q^{2} + (k+i-3)pq]}{p^{k-1}\,[1 + (k-3)q]^{2}}$$

$$\approx \frac{(k-2) + p + 4\,p^{2}/q}{p^{k-1}[1 + (k-3)q]^{2}}$$

for large k. The mean recurrence time for the "open" case is

$$\mu[k;\ k\text{-}1,\ k) = \frac{1 + (k-2)q + \sum_{i=0}^{k-2} p^{i}\,[p + iq + i^{2}q^{2} + (k-2)\,pq]}{p^{k-1}\,[1 + (k-2)q]^{2}}$$

$$\approx \frac{(k-2) + (1 + p)^{2}/q}{p^{k-1}[1 + (k-2)q]^{2}}$$

From the truncated sequence, the new estimate of p, the probability of a mutable site, is 34/148 = 0.2297. Using this value the expected recurrence times between events for k nucleotides (k = 3, 4, 5) were computed. The events were runs of k mutable sites, using the formula

$$\mu[k;\ k] = q^{-1}\,p^{-k} - q^{-1}$$

and then events defined as closed and open sets of k or k-1 mutable sites. These expected times (μ) were then converted to the expected number of occurrences of such events in a sequence of 148 nucleotides as 148 μ^{-1}. The figures are shown in Table 8 in the column headed E and compared with observed numbers (0) from Figure 1. Thus, the expected number of nonoverlapping occurrences of the event (MMM, MCM) in this sequence was 6.1, whereas for the event (MMM, MCM, MMC) is 9.5. It is clear that there is no evidence of hot spots of adjacent mutable sites in the sequence. The distribution of mutable sites accords with random expectations, once the region 131-158 is removed from consideration. When this modification is made to the distribution in Table 7, the observed variance becomes 14.48, which conforms closely to the value of 14.60 expected for the fitted geometric distribution.

Table 8. Frequency of clusters of mutable sites.

r	Run [k; k]		Closed [k; k, k-1]		Open [k; k, k-1)	
	O	E	O	E	O	E
3	1	1.4	9	6.1	10	9.5
4	1	0.3	4	2.1	4	3.0
5	0	0.1	0	0.8	1	0.9

These analytical methods could be extended to more complex sets of events, and the means and variances derived from differentiation of the relevant generating functions for recurrence times, as outlined by Feller.

F. DISTRIBUTION OF MULTIPLE CHANGES PER SITE

Besides indicating the distribution of nucleotide sites at which substitutions were observed, Figure 1 also shows the observed number of variant bases at each mutable site. In all but one of the positions (59), only one kind of variant nucleotide was found. It seems unlikely that multiple occurrences of the same variant nucleotide at a given site result from multiple independent mutational events.

To test this null hypothesis of independent mutations more formally, we construct a binomial model under the null hypothesis and derive expected frequencies. Each site is first classified into the number of copies which exhibited a mutant at that site. The basic random variable ranges from 0 to 5 mutants. (The range is limited by the definition of the consensus nucleotide as being the most frequent one at that position.) The site is then classified according to the number of kinds of variant nucleotides found at that site and, if more than one, their distribution (*e.g.*, 2A, 1G, 1C). As there is only one site in this example at which more than one kind of variant was found, it is sufficient for these data to distinguish only whether the multiple variants are the same nucleotide or not.

Under the null hypothesis which ascribes all variants to independent mutations, let m denote the probability of a single variant. We assume then that m is constant over all sites, and that there is no bias in nucleotide substituted, all three events being equally likely. The proportions of sites expected to show no mutations in 12 copies is $(1-m)^{12}$. The expected proportion of sites having changes in two of the copies is $m^2(1-m)^{10}\ 12!/(2!10!)$. Of those with two variant copies, one third would have the same variant nucleotide, and two-thirds would have two different nucleotides, one instance of each. In the case of sites with three variants, one-ninth would show the same variant nucleotide and eight-ninths would be mixed. Likewise, for four variants, 1/27 would be the same and 26/27 would be mixed. The observed number of nucleotide sites so classified is given in Table 9.

The maximum likelihood estimate of m is 51/2112 = 0.0241. Using this estimate, the expected number of sites in each category was computed and is also shown in Table 9. The value of chi-square for these expectations is 67.26 ($P < 0.005$), so the null hypothesis is rejected.

In section 3 above, we established that the probability of mutation was not uniform with respect to the kind of base subject to change or the kind of base introduced. Furthermore, in section 4 a sector of 27 bases was found to be significantly conserved. Such distortions would affect the above expectations. Therefore, the expectations were recomputed using the observed 4 x 4 incomplete contingency table (Table 2) and deleting those bases in the conserved region. Separate expectations were computed for each of the four kinds of bases, for each category of Table 9, and summed to give the total expectation. Thus, for example, at the (38 - 1) x 12 original cytosine positions of the 12 sequences, a total of 19 variants were found (5G, 13T, 1A). The estimate of m for cytosine was 19/(37 x 12) = 0.0428. The expected number of cytosine sites showing two identical variants is

$$37 \times 66 \times (0.9572)^{10}\ (0.0428)^2\ (5^2 + 13^2 + 1^2) / 19^2 = 1.5346$$

Table 9. Distribution of site numbers according to variants found at site.

Number of Variant Copies for Each Site		Number of Nucleotide Sites			
		Observed Number	Expected Proportion*	Expected Number	Expected No. Corrected
0		141	$(1-m)^{12}$	131.25	107.00
1		22	$C_1\ m\ (1-m)^{11}$	38.98	34.18
2	Identical	10	$(1/3)C_2\ m^2(1-m)^{10}$	1.77	4.34
	Mixed	1	$(2/3)C_2\ m^2(1-m)^{10}$	3.54	2.43
3	Identical	1	$(1/9)C_3\ m^3(1-m)^9$	0.05	0.46
	Mixed	-	$(8/9)C_3\ m^3(1-m)^9$	0.39	0.49
4	Identical	1	$(1/27)C_4\ m^4(1-m)^8$	0.00	0.04
	Mixed	-	$(26/27)C_4\ m^4(1-m)^8$	0.02	0.06

Class		Chi-square Summary		Corrected	
		Observed	Expected	Observed	Expected
0		141	131.25	114	107.00
1		22	38.98	22	34.18
over 1	Identical	12	1.82	12	4.84
	Mixed	1	3.95	1	2.98
	$\chi^2_{(2)}$		67.26		16.71

*Where C_i is the combinatorial coefficient 12!/[i!(12-i)!].

The corrected expectations are shown in the last column of Table 9. The adjusted chi-square value of 16.71 indicates a much closer fit, but the assumption of independence must still be rejected ($P < 0.005$). Thus, part of the discrepancy in the pattern of distribution of multiple changes at a site is due to varying substitution rates among the bases replaced, and the bases substituted, and the restriction of substitutions away from the conserved sector. However, a large part of the discrepancy is due to the possibility that the observed distribution reflects the outcome of *two* distinct processes: mutation, and the propagation of a single mutant event to other copies of the sequence ("concerted evolution").

VI. RELATIONSHIP AMONG SEQUENCES

In the preceding section, it was concluded that a significant fraction of the multiple occurrences of the same variant nucleotide at a particular site was due to sharing of the same mutational event. This implies that two particular copies which share the same substitution are more closely related than any two of the 12 copies on average. It follows that such data reflect the order of mutational events separating different copies of the basic sequence.

Table 10 summarizes the distribution of shared variant nucleotides over the samples of 12 copies. The copies are labelled A, B, ..., K where A denotes the consensus sequence which was recovered twice in the sample of 12 copies. The alphabetical order of these letters corresponds to the rank of the total number of mutational changes in the sequence compared with A. Copy B differs from the consensus sequence A, but the difference is not at one of the relevant positions.

A tree expressing the relationships among the copies based only on the shared mutations can be constructed. In this tree, each segment of a branch corresponds to a mutation at any of the 12 sites under consideration. Sectors of the branch stemming back to A, B and shared between two sequences indicate shared mutations. The aim in tree construction is to maximize the number of shared internodes and minimize the total of the number of mutations at shared sites separating each terminal sequence from its last node (see chapter by Felsenstein in this volume for a more complete account). Figure 2 diagrams such a tree, in which the total number of terminating private mutations required to generate the sequences is ten. The implication of this tree is that multiple occurrences at sites 40, 60, 111, 127, 129, 167 and 173 are explained by derivative relations among the sequences. However, for each of the five sites 22, 53, 58, 75 and 115, two independent events must be invoked. The possibilities are (1) identical substitution occurred twice at same position, and (2) a single mutation, but recombination between the sequences from different lineages. Possibility (1) is reasonable in view of section V in which biased rates of substitution were corrected.

Table 10. Position of variants shared by more than one copy.

Copy	Position of Shared Variant											
	22	40	53	58	60	75	111	115	127	129	167	173
C											A	
E						T					A	
D	T											C
F			A									C
I								A	C	A		C
H				T		T			C	A		C
G		T						A				
J		T	A		A		T					
K	T	T		T	A		T					

The data are too few to determine whether or not some recombination between sequences has taken place. We note that the corrected expectations for the number of sites with identical double mutants is 4.34 (Table 9). This number accords with that indicated in the tree of Figure 2 so that there is no need to postulate recombination.

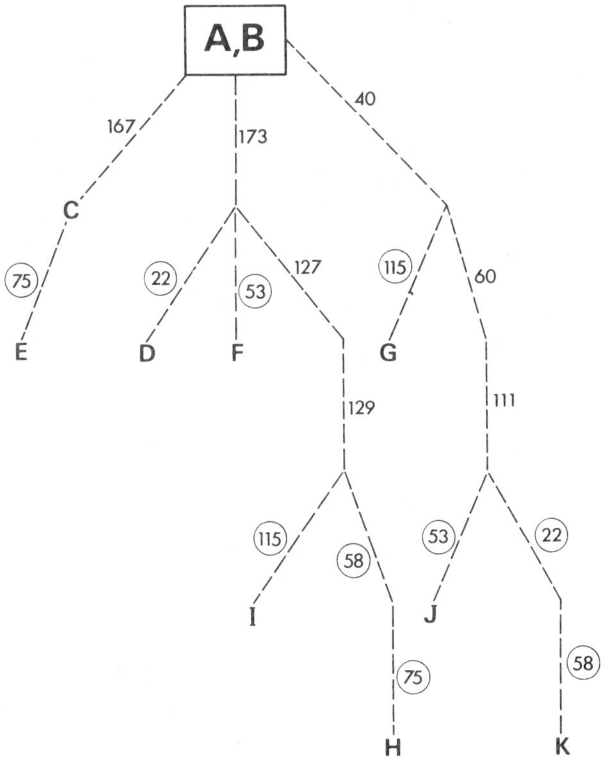

Figure 2. Network of relationships among the copies (A, B, ..., K) based on shared variants at particular sites (Table 10). The positions of the seven sites of shared variants used in this network are shown as numbers, whereas the other five sites with shared variants are shown as numbers enclosed by circles.

The topic of sequence relationships is, of course, much broader than is dealt with here, with an extensive treatment given by Felsenstein in this volume.

VII. JOINT DISTRIBUTION OF MUTANTS OVER SEQUENCES

The final analytical technique to be discussed here is concerned with the correlation structure of variation at all the sites. In this case, we regard variable nucleotide sites in a sample of copies as polymorphic loci in a sample of haploid genomes. In the traditional terms of population genetics, the concern is with the extent of linkage disequilibrium, or of multilocus structure.

There are several ways in which association between variants might be measured and summarized (Karlin and Piazza, 1981). One recent proposal (Brown *et al.*, 1980) which is useful in sequence data is based on the distribution f(K) of the number of dissimilar (heterozygous) sites (K) when two randomly chosen copies (gametes) are compared nucleotide by nucleotide.

In the present example of 12 copies there are 66 such comparisons. The observed number of comparisons in which two copies (sampling without replacement) differ at K = 0, 1, 2, ... nucleotide sites is given in Table 11. The mean of this empirical distribution is predictable solely from the diversity at each of the 176 sites. The diversity h_i is the probability that two randomly chosen copies will differ at a single particular site, where i is an index referring to a class of the sites with the same diversity value. Thus, $n_1 = 141$ of the 176 sites had no variants among the 12 sequences, and all have $h_1 = 0$. A second class of $n_2 = 22$ sites had one variant in 12 sequences and so had a diversity $h_2 = 1/6$. One site, $n_3 = 1$, had two different variants and 10 with the consensus nucleotide, and so had a diversity of $h_3 = 7/22$. Ten sites, $n_4 = 10$, had 10 consensus and two identical variants, so that $h_4 = 10/33$. The fifth and sixth classes were represented by one site each, with three and four identical variants, $n_5 = 1$, $h_5 = 9/22$; $n_6 = 1$ and $h_6 = 16/33$. Hence, the mean of the observed distribution in Table 11 is

$$\bar{K} = \sum_{i=1}^{6} n_i h_i = 7.91$$

The higher moments of the distribution f(K) depend not only on h_i but also on the correlation among the variants at different sites over the sequences.

Table 11. Observed and expected distribution of the number of dissimilar (heterozygous) sites when two copies are compared.

Number of Dissimilar Sites	Number of Comparisons	
	Observed	Expected
0	1	0.01
1	5	0.06
2	5	0.32
3	6	1.12
4	3	2.77
5	3	5.30
6	2	8.10
7	6	10.18
8	7	10.72
9	4	9.61
10	3	7.40
11	4	4.45
12	6	2.89
13	3	1.49
14	1	0.67
Variances	$s_K^2 = 23.33$	$\sigma_K^2 = 5.93$

Let us consider the null hypothesis that there is no association among variants - that is, that occurrence of a variant nucleotide at one position of the sequence is independent of whether a variant occurs at other sites. Under this hypothesis, the expected distribution f(K) can be obtained as the coefficients of powers of x in the generating function G(x), which is a convolution of the generating functions of the separate independent Bernoulli distributions for each site.

$$G(x) = \prod_{i=1}^{6} (1 - h_i + h_i x)^{n_i}$$

These coefficients were obtained and multiplied by 66 to arrive at the expected number of comparisons with various numbers of heterozygous sites. The values are given in Table 11. It is clear that the observed distribution differs markedly from the expected in that the observed is much more dispersed. Indeed, the observed variance s_K^2 is 23.33 whereas the expected variance under the null hypothesis is (Brown *et al.*, 1980)

$$\sigma_K^2 = \Sigma n_i h_i - \Sigma n_i h_i^2 = 5.9$$

There is a fourfold inflation of the variance due to the association of variants at different sites. Unfortunately, the appropriate testing theory for small samples has yet to be developed. Nevertheless, the statistic $[s_K^2/\sigma_K^2] - 1$ is a convenient index of mutilocus structure for comparisons to other sets of data. Here the large departure from zero again suggests that concerted evolution among copies of this sequence has taken place.

VIII. CONCLUSIONS

From the statistical standpoint, sequence data differ markedly from data previously available to geneticists (such as allozymes or morphological polymorphisms). Sequence data usually consist of only a few copies. Each copy, however, will normally include many nucleotide sites. To illustrate this point, suppose that each nucleotide site is computationally analogous to an allozyme locus. In the typical allozyme study of plant populations (see Hamrick *et al.*, 1979), an average of 12 loci are surveyed per individual. Such loci are usually scattered throughout the genome. In contrast, on a single DNA sequencing gel, about 250 sites can be read (see photograph in chapter by Gingeras in this volume). These sites are adjacent and so are highly correlated. From the standpoint of numbers of organisms, however, allozyme surveys readily cope with sample sizes of 100, and studies based on over 10,000 individuals per population have been published. It is unlikely that sample size will exceed 20 in DNA sequencing so that, statistically speaking, small sample effects must be taken into account. The full generality of statistical conclusions based on the typical limited data set can be tested only as the results of several independent studies are accumulated.

Despite this limitation, sequence data are rich in empirical content and afford an unparalleled view of the basic processes of genetic change (Clegg, 1982). Moreover, the complete analysis of these data poses a number of practical statistical problems.

The present set of data has served to illustrate some of the questions which arise naturally from DNA sequences and their associated statistical analyses. We have seen that the pattern of nucleotide substitution contains a great deal of structure in the sense that the elementary hypothesis of random substitutions is inadequate.

First, there is a marked bias in the substitution of nucleotides. This bias is reflected in both the probability that a given nucleotide will be replaced [*e.g.*, Prob$(A \rightarrow T) \neq$ Prob$(G \rightarrow T)$] and also in the probability that a given nucleotide will be introduced [*e.g.*, Prob$(G \rightarrow A) \neq$ Prob$(G \rightarrow T)$].

Second, the distribution of variable sites along the molecule is nonrandom. One interval of 27 bp is highly conserved. Differential conservation may indicate selective constraints associated with functional properties. Indeed, the analysis of DNA sequence variation may become an important tool in the search for functional properties.

Third, the distribution of mutant events at a given site does not accord with the assumption of independence. Evidently the same mutant events are sometimes propagated to other repeating units in the multigene family. While the mechanism of propagation is unknown, the phenomenon which has been termed concerted evolution (Arnheim *et al.*, 1980) has been described in other systems.

The fact that genes at different loci exchange mutant sites and therefore evolve in a highly correlated fashion conflicts with time-honored assumptions in population genetic theory. For instance, calculations on the cost of evolution, which assume independence among loci, are now suspect. Further evidence for concerted evolution comes from the analysis of the joint distribution of mutant sites among copies. Here again, the pattern is one of strong correlation among mutant sites. Finally, it is possible to derive networks of relationships among copies based upon the shared distribution of mutational sites. As data accumulate, it should become possible to estimate the rate of genetic exchange among repeating units and to use rate estimates as a basis for dating the separation of different lineages. A host of fascinating problems in evolutionary genetics are likely to yield new insights when posed in the context of DNA sequence information.

ACKNOWLEDGMENTS

We are very grateful to Dr. E. S. Dennis and Dr. W. J. Peacock for making available to us the sequence data of the clones of knob heterochromatin and for encouragement and criticism of this chapter. This work was conducted while M. T. Clegg held a John Simon Guggenheim Foundation Fellowship and was a visiting scientist at the CSIRO Division of Plant Industry.

Chapter 6

INFERRING EVOLUTIONARY TREES FROM DNA SEQUENCES

Joseph Felsenstein

Department of Genetics SK-50, University of Washington, Seattle, Washington

I. INTRODUCTION

If we were to know the true genealogy of all life, draw a diagram of it, and look at that from a sufficient distance, the details of individual matings would blur into single lines for each species and the whole would appear as a great branching tree. There are exceptions to this pattern, particularly in bacteria and plants, but as a first approximation we may take it to be a rooted tree with a geological time scale on which the nodes of the tree can be placed.

Evolutionary trees are portions of this tree. They are usually called phylogenies, although when the topology of the tree is provided without any time scale it is often called a "cladogram". Phylogenies sometimes also contain information about the phenotypes inferred for the ancestors. It is necessary in each case to be clear about what is meant by a phylogeny.

Molecular data, particularly nucleic acid and protein sequences, provide us with perhaps the best source of information on phylogenies. While it is extremely difficult to find morphological characters which allow us to compare yeast and humans, molecular sequences allow this, and the processes governing their evolution seem comparable over wide taxonomic ranges.

It was recognized very early, by Pauling and Zuckerkandl (1963), that a "chemical paleogenetics" was possible. The first evolutionary trees obtained from protein sequences were done by methods not explicitly stated. Fitch and Margoliash (1967) presented a reasonably well-defined method for inferring phylogenies based on a matrix of percentage differences between sequences. This has since largely been replaced by the "ancestral sequence method" of Eck and Dayhoff (1966), which is usually referred to as the "maximum parsimony" method.

While the parsimony method is not derived from statistical considerations or stated in statistical terms, no treatment of the inference of phylogenies can ignore it, since it is by far the most common method used by molecular evolutionists. What is known of its properties is not entirely encouraging.

II. PARSIMONY METHODS

A. The Parsimony Criterion

Parsimony methods originated with Edwards and Cavalli-Sforza (1963, 1964) who were analyzing gene frequencies and adopted the approach only when they could not get maximum likelihood to work. Camin and Sokal (1965) introduced them for morphological characters coded into discrete states. Their use in molecular evolution was initiated by Eck and Dayhoff (1966).

For each phylogeny, we can calculate the smallest number of nucleotide substitutions which could be invoked to explain evolution of the observed data on that phylogeny. The parsimony criterion directs us to prefer that phylogeny which requires the fewest substitutions. Edwards and Cavalli-Sforza (1963) gave it the more precise name of the "method of minimum evolution". In the case of nucleotide sequence data we allow substitution of any base for any other. This has the effect of removing all information as to the location of the root of the tree.

Figure 1 illustrates this. Four observed nucleotide sequences are shown, connected by an unrooted tree. Given this particular unrooted tree, it is possible to explain the evolution of these observed sequences with as few as 12 substitutions. In parentheses at the interior nodes of the tree are shown two hypothetical sequences of those ancestors. For each edge of the tree (each line) we can compute the number of substitutions required to go from the sequence at one end to the sequence at the other. In this case the sum of these is 12. There are other possible assignments of the interior node sequences which achieve the same sum.

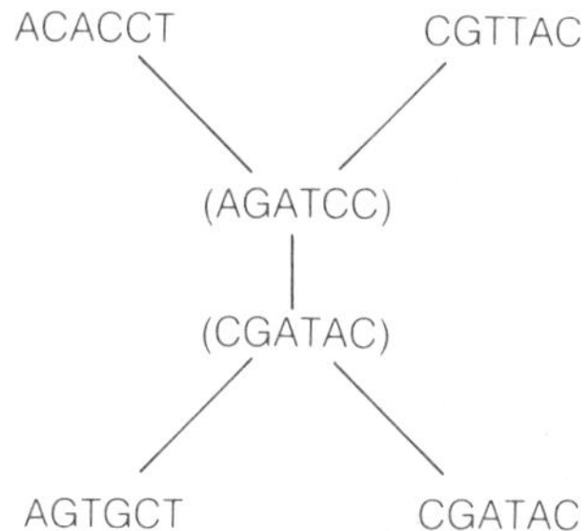

Figure 1. An example of nucleotide sequences in four species connected by one of the three possible unrooted tree topologies. Hypothetical ancestral sequences (in parentheses) are shown for the unobserved nodes of the tree, which requires 12 nucleotide substitutions, the minimum possible under this topology.

Note that we could root the tree anywhere and have the same total number of substitutions. The root could be placed at any of the six nodes shown, or on any of the five edges shown. If the root is placed at the upper leftmost sequence, the first three changes on this tree are C $\rightarrow$ G, C $\rightarrow$ T and T $\rightarrow$ C. If the root were instead placed at the uppermost interior node, the changes along that edge are instead inferred to be G $\rightarrow$ C, T $\rightarrow$ C and C $\rightarrow$ T, but their number is not affected. The parsimony method thus infers not a rooted tree but an unrooted one, representing an equivalence class of rooted trees, all of which have equal numbers of substitutions.

It has become conventional in systematics to call an unrooted tree a "network", although this conflicts with standard mathematical usage in graph theory. The tree can be rooted by external considerations, including acceptance of the notion of a "molecular clock", according to which we expect equal amounts of change in a sequence per unit time along different lineages.

Efficient algorithms exist which allow us to compute for any given unrooted tree the minimum number of substitutions needed along that tree. For nucleic acid sequences the method is due to Fitch (1971). Hartigan (1973) and Sankoff (1975) have generalized Fitch's method. For protein sequences the problem is much more difficult, if one is trying to find the smallest number of nucleotide substitutions which could have led to the given protein sequences. Moore *et al.* (1973) gave a complicated algorithm which achieves this goal: Fitch and Farris (1974) gave a quicker approximate method which may not always give the correct answer.

Fitch's (1971) method starts by rooting the tree at an arbitrary point. It then considers the sites one at a time. For each site, it constructs at each tip (terminal node) of the tree the set of nucleotides which could possibly exist at that point. We then move down the tree (in computer science terminology, we traverse it in postorder), filling sets of nucleotides at the interior nodes and counting steps by the following rules:

(1) If the two sets immediately above a node are disjoint, count one step and place their union at the node.
(2) If they are not disjoint, count no steps and place their intersection at the node.

Figure 2 shows the tree of Figure 1, rooted at a point on the lower leftmost edge, with Fitch's algorithm carried out for the first site. Note that, although the algorithm computes the number of steps correctly, it does not make an explicit reconstruction of sequences at the interior nodes.

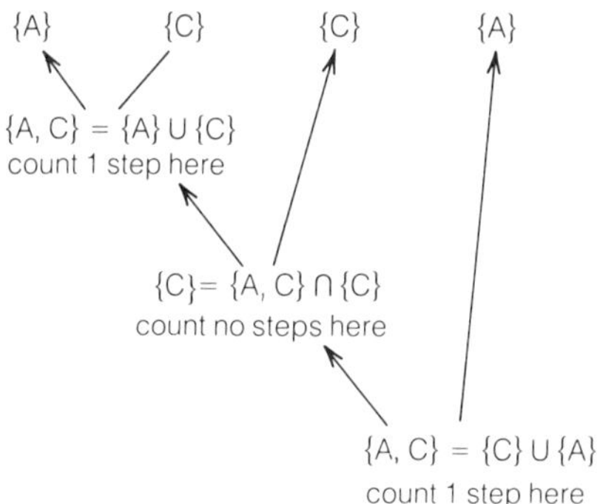

Figure 2. Fitch's (1971) algorithm, applied to the first nucleotide of the data and tree from Figure 1. That tree has been rooted on the lower leftmost segment.

B. Finding the Most Parsimonious Tree

The tree in Figure 1 is not the most parsimonious tree. Figure 3 shows the most parsimonious tree, and it has a different topology and requires only eight substitutions. Since there are only a finite number of possible trees (in this case three), we could, in principle, search for the most parsimonious tree by considering all possible unrooted trees and counting the minimum number of substitutions needed for each one. The large number of possible trees (over 10^{20} unrooted trees for 20 species; Cavalli-Sforza and Edwards, 1967; Felsenstein, 1978a) makes this an impracticable method in most cases.

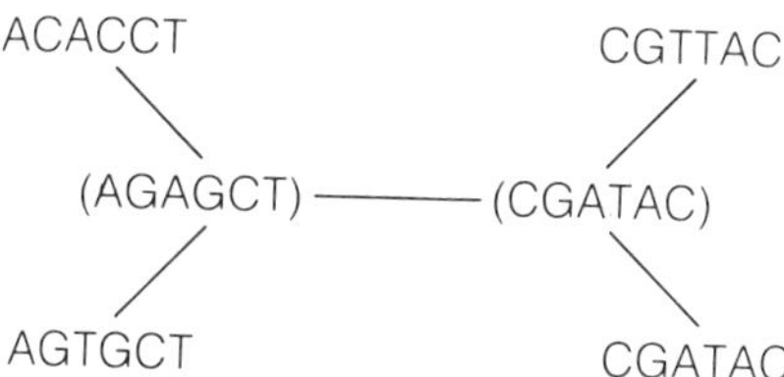

Figure 3. The unrooted tree requiring the fewest nucleotide substitutions for the data used in Figure 1. This is the "most parsimonious" tree. Hypothetical ancestral sequences have been reconstructed consistent with the minimum number of nucleotide substitutions, although there are other possible ancestral sequences which also require only eight substitutions.

We may consider a space consisting of all possible sequences (so that this space will have 4^{40} points in it for sequences of 40 nucleotides). We construct a graph by connecting adjacent sequences: we consider two sequences adjacent if they differ by only one substitution. This is a difficult graph to visualize: it has $4^{40} \approx 10^{24}$ points, no two of which are more than 40 steps apart, for sequences of length 40. Our data consist of a set S of these points, the observed sequences. The most parsimonious trees (there may be more than one) correspond to Steiner Trees in this graph: subgraphs consisting of the set S and a set R of "Steiner points" which together form a tree of minimum length.

Finding a Steiner Tree in an arbitrary graph is a problem which is known to be NP-complete (Karp, 1972). Foulds and Graham (1982) show that, in the graph generated by sequences using only two nucleotides, the Steiner Tree problem is NP-complete, so that with four nucleotides it is NP-hard. This, in effect, means that we cannot find a method for solving the Steiner Tree problem for nucleotide sequences, which is guaranteed to work in a time polynomial in the number of species.

More practical methods work by starting with an initial rough guess at the tree, or by adding species one at a time to the tree, and in either case doing repeated local rearrangements of the tree, accepting those which improve the tree. These methods are not guaranteed to find a Steiner Tree, nor will they permit us to know when we have found one. The most they can guarantee is that no local rearrangement will improve the tree over the one found.

Fitch (1975, 1977) and Penny (1982b) have proposed branch-and-bound methods which economize on the search for Steiner Trees, and which can guarantee finding them all, though not necessarily in polynomial time. Hendy *et al.* (1978) and Foulds *et al.* (1979) give a method which, if it succeeds, will allow us to know that a Steiner Tree has been found, but it will not always succeed.

The parsimony method is a very seductive one: its very name seems to promise logical soundness to the scientist, and mathematicians are attracted by the respectability of the combinatorial optimization problems it generates. Here the statistician has a critical role to play, for it turns out that there is little immediate connection between the parsimony method and statistical criteria, and a statistical viewpoint casts a harsh and revealing light on contemporary phylogenetic practice.

C. Statistical Properties of Parsimony

1. Is Parsimony Maximum Likelihood? One of the earliest attempts to provide a statistical foundation for the use of the parsimony method was Farris's (1973) attempt to show that parsimony makes, in effect, a maximum likelihood estimate. Farris argued that, if one were estimating an "evolutionary hypothesis" which consisted of a phylogeny plus estimated ancestral node sequences, a simple mathematical argument proved that the maximum likelihood estimate always had the same tree topology as the parsimony estimate. Given the premise, Farris's argument is convincing, but the question arises as to whether this is a satisfactory model for estimation of the phylogeny. I demonstrated (Felsenstein, 1973) that when we estimate not the "evolutionary hypothesis" but the phylogeny, including node times but not hypothetical ancestral sequences, the phylogeny obtained by maximum likelihood was not always the same as that obtained by maximum parsimony.

I was able (Felsenstein, 1973) to show that parsimony and maximum likelihood do become the same when rates of change become very small, so that, *a priori*, we expect to see little difference between the sequences. This makes intuitive sense: if we expect to see little change, it is reasonable that the appropriate statistical method will try to bring expectation into line with observation by finding that tree which requires as few as possible of the unsettling events. More detailed discussion of this matter will also be found in my later paper (Felsenstein, 1978b).

It is interesting that the maximum likelihood estimate of the "evolutionary hypothesis" does not yield the same tree topology as the estimate of the phylogeny. One possible source of the discrepancy is the presence of ancillary parameters (the ancestral sequences) in the estimation of the "evolutionary hypothesis". As the length of the sequences is increased, the number of these ancillary parameters increases without limit. In the estimation of the phylogeny, the only parameters which may be considered ancillary are the node times, which do not proliferate as longer sequences are taken.

2. Parsimony and Consistency. The other approach to statistical analysis of the parsimony method is to take it as given and analyze its properties directly. I have looked into the consistency of the parsimony estimate (Felsenstein, 1978b), on the assumption that consistency is the most desirable property an estimator can have. Although my argument assumes morphological characters with two states (0 and 1), it generalizes naturally to the case of four bases.

We assume a simple stochastic process of base substitution along the tree, which is assumed to operate independently at each site. The probability that a given site changes in segment i of the tree is taken to be the arbitrary quantity P_i, with each of the three possible changes (*e.g.*, A → C, A → G and A → T) accounting for one-third of this probability. The fact that, under this model, probabilities of change are not necessarily related to length of the tree segment in time allows for models which do not incorporate a "molecular clock".

Consider as an example the four-species tree shown in Figure 4, whose values of P_i are of two sizes, p and q. The only outcomes at a site which affect the results of the parsimony method are, for a four-species tree, those in which two different bases are present, each represented twice. All other outcomes will require the same number of substitutions on all three tree topologies. There are three possible patterns in those characters which affect the tree topology: C G G C, C G C G and C C G G. Only the last favors the correct topology shown in Figure 4. We can compute the probabilities of the three patterns and they are shown in Table 1.

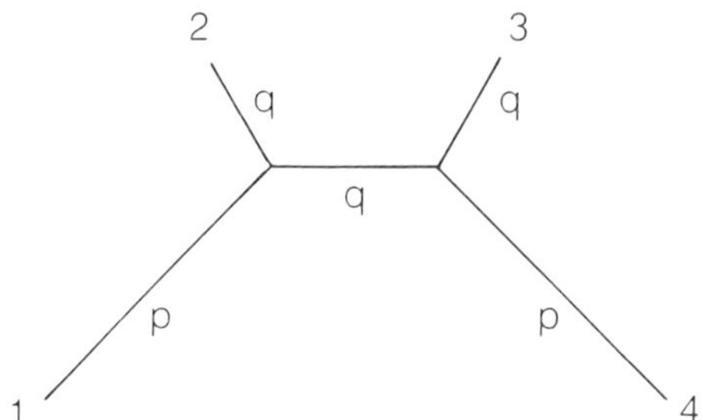

Figure 4. An unrooted tree with probabilities of nucleotide substitution beside each segment. When p is large and q small, the parsimony method will not make a consistent estimate of this tree.

A simple argument using a law of large numbers shows that, as we take more and more sites (longer and longer sequences, or more and more of them for the same species), the numbers of sites falling into these three classes will converge to their expected frequencies. Which of the three possible tree topologies results from application of the parsimony method will then depend on which of these three outcomes is the most probable. When we take q small and p large (it cannot be larger than 3/4 if it is to be consistent

Table 1. Probabilities of patterns for four-species trees.

Pattern	Probability
x x y y	$(81q - 108pq - 162q^2 + 27p^2q + 252pq^2 + 81q^3 - 144pq^3 - 84p^2q^2 + 64p^2q^3) / 81$
x y y x	$(27p^2 + 27q^2 - 27q^3 - 36pq^2 - 81p^2q + 48pq^3 + 108p^2q^2 - 64p^2q^3) / 81$
x y x y	$(54pq - 45p^2q - 108pq^2 + 9q^3 + 108p^2q^2 + 48pq^3 - 64p^2q^3) / 81$

with a continuous-time symmetric Markov process of base substitution), the three outcomes approach probabilities q, 3/16 and q/2. Since q is small this establishes that the parsimony result is not, in this case, a consistent estimator of the phylogeny.

The pattern which emerges is that the more unequal are p and q, the less likely is the estimate to be consistent. The smaller are p and q, the more unequal they must be to result in lack of consistency. In this sense the result does not contradict the proof that the parsimony method gives a maximum likelihood estimate if rates of change are small. Note that the above case departs rather far from the molecular clock assumption: when the molecular clock holds, it may well be that parsimony estimates are always consistent, though no general proof to this effect has yet been produced.

3. Confidence Regions for Parsimony Analyses. How can we associate a statistical test with the parsimony method? Cavender (1978, 1981) has pioneered investigation of this question. He also confined his attention to two-state morphological characters, but his model is also easily generalized to four bases. In fact, it can be taken to be the same model as discussed above, and in turns out that the critical case is the one shown in Figure 4.

Cavender asked what set of probabilities of change P_i for one topology gave the strongest evidence favoring another topology. The answer was the case shown in Figure 4 where (when p is 3/4 and q is zero) 3/16 of the characters favor one particular incorrect topology by one substitution and none favor the correct one. Given this result, the critical region for rejecting the topology shown in Figure 4 turns out to be simply the tail of a binomial distribution with parameters n (the number of sites) and 3/16. Since it is different from Cavender's original table, Table 2 shows the critical points, giving the number of substitutions C by which one tree must exceed another before the less parsimonious tree can be rejected.

Table 2. Number of substitutions C by which one tree must exceed another before less parsimonious tree can be rejected with n sites.

n	$C_{0.95}$	$C_{0.99}$	n	$C_{0.95}$	$C_{0.99}$
1	-	-	20	8	9
2	2	-	25	9	11
3	3	3	30	10	12
4	3	4	40	13	15
5	3	4	50	15	17
10	5	6	75	21	23
15	6	8	100	26	29

The sizes of the numbers in this table are striking and must give pause to those who become obsessed with finding the most parsimonious tree, even when that tree is only slightly better than the next best tree. The critical number of extra steps is always at least 3/16 of the number of base positions.

Cavender's results hold only for four species. Nothing has yet been done to extend it to five or more species, or to make a serious investigation of how it is affected by the assumption of a molecular clock.

III. MAXIMUM LIKELIHOOD METHODS

The direct use of maximum likelihood methods to estimate evolutionary trees from DNA sequences was initiated by Neyman (1971). Kashyap and Subas (1974) extended Neyman's approach by combining estimates from triples of species. These authors were not able to obtain maximum likelihood estimates for more than three species. Kaplan and Langley (1979) and Kaplan and Risko (1981) have developed maximum likelihood methods for estimating divergence time of two species from restriction site data. We are concerned here with complete sequences.

I have re-examined the matter recently (Felsenstein, 1981), showing how economies can be made in the computation and maximization of the likelihood. One can start, like Neyman, with a stationary stochastic process of base substitution which has transition probabilities from state i to state j in a lineage over time t of $P_{ij}(t)$, the states taking the values A, C, G and T. We assume that each site evolves independently, and that lineages, once separated, evolve independently. We also assume that the equilibrium distribution of the four bases under this process is known and is given by p_A, p_C, p_G and p_T.

A. Computing the Likelihood

Figure 5 shows a four-species rooted tree whose likelihood we will compute. The likelihood is the product of probabilities of independent events occurring in different sites. Suppose that we denote the base found at node i in position j by $s_i(j)$. For the given tree the likelihood will be

$$L = \prod_{i=1}^{n} \sum_{j}\sum_{k}\sum_{\ell} p_j P_{j,k}(v_5) P_{j,\ell}(v_6) P_{k,s_1(i)}(v_1) \times P_{k,s_2(i)}(v_2) P_{\ell,s_3(i)}(v_3) P_{\ell,s_4(i)}(v_4) \tag{1}$$

the summation over j, k and ℓ being over the four bases, and the lengths of the tree segments v_i being indexed by the number of the node at the upper end of the segment. The observed sequences affect the likelihood through the $s_i(j)$, and the tree affects the likelihood through the particular P's in the product and the lengths v_j.

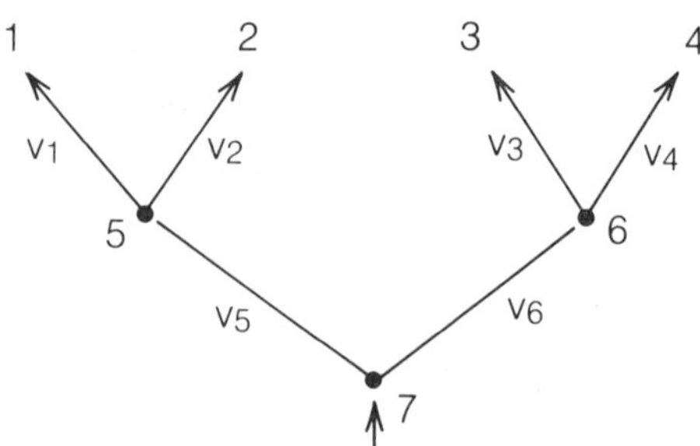

Figure 5. The rooted tree for which the likelihood is given by equation (1).

For trees of even moderate size the summations, which are over all possible combinations of states at interior nodes, become extremely onerous. Fortunately the summation signs in (1) can be moved rightwards and the expression factored in a way similar to Horner's Rule for evaluating polynomials. The resulting expression has a pattern of parentheses exactly following the structure of the tree.

That, in turn, leads to an algorithm for evaluating the likelihood by moving down the tree (traversing it in postorder). It computes at each node a quantity $L^{(j)}(s_j)$, the likelihood of that part of the tree at or above node j, conditional on node j having base s_j at the site in question. When the bottom of the tree is reached, the likelihood of the tree at the site can be computed by combining the four quantities at that node, each weighted by the base frequency p. The interested reader will find the equations in my paper (Felsenstein, 1981).

B. Maximizing the Likelihood

The ability to compute the likelihood rapidly still leaves us with the problem of finding the tree and the segment lengths which maximize the likelihood. This is enormously facilitated if the stochastic process $P_{ij}(t)$ is reversible, as the processes used by Neyman (1971), by Kaplan and Langley (1979) and by me have been. The particular process I have used is the continuous-time Markov process for which

$$P_{ij}(dt) = (1-dt)\delta_{ij} + dt\ p_j \tag{2}$$

The reversibility turns out to guarantee that the root of the tree cannot be inferred: in the tree shown in Figure 5 the likelihood will depend on v_5 and v_6 only through their sum $v_5 + v_6$. If we do not assume a molecular clock, allowing the v's to vary independently of one another, it turns out that there is a way to maximize the likelihood with respect to the v_i one at a time. The resulting algorithm is related to the EM algorithm for maximizing likelihoods

(Dempster *et al.*, 1977). Although it guarantees only achievement of a stationary point on the likelihood surface, it seems to do well in practice.

As a result of all this, we have an algorithm for finding a maximum likelihood for a given unrooted tree topology, but we still must search among topologies. The asymptotic covariance matrix of the segment lengths can be computed, though only with much effort. It is more economical to compute its diagonal elements only, using them to get a rough idea of the error. Even then, the resulting computer program is extremely slow. (The program is available in the phylogeny inference package PHYLIP, a package of Pascal programs which I will supply free on a magnetic tape supplied by the recipient. The package also includes programs for carrying out parsimony and compatibility methods on nucleic acid sequence data.)

C. Limitations of the Model

Maximum likelihood methods will in this case have favorable properties: a version of the argument given by Rao (1965, pp. 295-296) can be used to establish their consistency. Nevertheless there is much room for improvement in maximum likelihood methods for inferring phylogenies from DNA and RNA sequences. The heavy computational burden places a premium on finding ways to speed up computations. But the model itself is also overly simple and needs considerable development. There has already been some development of models of transition among bases which take account of the greater frequency of transitions as opposed to transversions (Kimura, 1980, 1981; Takahata and Kimura, 1981), although these have not yet been incorporated into a likelihood analysis.

In the present model, the rates of change at all sites are equal. In coding regions, change is likely to be substantially more frequent in the third position than others, and more frequent in the first position than in the second. A full model would take into account which amino acids are coded for by each codon, and allow more frequent change to chemically similar amino acids. This would introduce a lack of independence at adjacent nucleotides.

It would also be desirable to correct for the avoidance of termination codons and to allow a limited amount of deletion and insertion, primarily in multiples of three bases. In addition, the codons should be imagined to be of two types: those coding for the active site and which only rarely can have a substitution allowed by natural selection, and those which can change more readily. It will often not be known in advance which is which. Noncoding sequences will presumably show no effects of position of the base in a codon, but much more deletion and insertion in addition to nucleotide substitution. However, little is known about the function of most of the DNA in the genome, and this must lend a roughness to any estimate and cause the computed error variances to be underestimates.

Attempts have been made to incorporate deletions and insertions into parsimony methods (Sankoff and Rousseau, 1975; Sankoff, 1975), but they have not yet been taken into account in a fully statistical method, which is not surprising since they introduce substantial nonindependence of sites.

We have repeatedly mentioned the molecular clock hypothesis in the above discussion. The molecular clock was originally predicted as a consequence of the neutral mutation theory and is at least approximately valid (Wilson *et al.*, 1977). However, recent population genetic theory (Gillespie and Langley, 1979) shows some departures from the clock to be predicted even within a neutral mutation hypothesis. Taking these predicted departures into account in tree estimation will be extremely difficult.

D. Hypothesis Testing

The possibility obviously exists of using maximum likelihood to construct tests of hypotheses using the likelihood ratio test. Various tests of unequal rates of evolution in different genes, different sites and between transitions and transversions are obviously possible.

The molecular clock hypothesis is also subject to a likelihood ratio test, if we compare the likelihood with v_i constrained to be proportional to time to the likelihood when they are not constrained. Unfortunately, computationally effective methods of maximizing the likelihood under this constraint are not yet available.

A test of the punctuated equilibrium theory, on the other hand, is not possible since unobserved speciations and extinctions along the lineages could cause that theory to be compatible with any conceivable pattern of molecular evolution. The proponents of that theory have, in any case, not suggested that it affects molecular evolution in any particular way.

Tests comparing a given topology to alternatives are more problematic: the theory of the likelihood ratio test is an asymptotic theory, and when the amount of data is large one is within a single topology. Figure 6 shows the problem. We imagine two neighboring topologies, each with a parameter (x or y). When either parameter is zero the tree has a trifurcation, a condition intermediate between these two topologies. If the dashed line indicates likelihoods which are within an allowable likelihood ratio of the maximum likelihood, are we to reject the second topology or accept it? Asymptotic theory provides no guide.

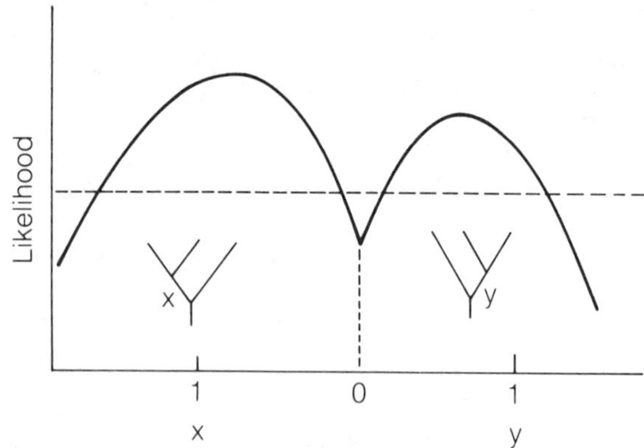

Figure 6. The likelihood plotted for a hypothetical case in which peaks are present in two adjacent topologies. Although the likelihood is acceptably high in the second topology, it is not in the trifurcating tree which connects these two tree topologies (the case where x = 0 or y = 0).

IV. ALTERNATIVES TO LIKELIHOOD

We do not have space here to examine in detail statistical approaches other than maximum likelihood. These have scarcely been developed. Carefully developed nonparametric approaches are obviously of interest. The main alternative framework currently in existence is the least squares methods (Fitch and Margoliash, 1967; Cavalli-Sforza and Edwards, 1967; Chakraborty, 1977). These reduce the data to a matrix of differences between sequences, attempting to fit the observed matrix to an expected one by least squares techniques. Chakraborty's paper is of particular interest in attempting to correct for some of the correlations among elements of the matrix, although it would need modification to be applied to DNA sequence data.

There has been no serious investigation of how much information is lost in reducing the data to differences. The approach may save considerable amounts of computation, though at what cost in efficiency of estimation is not known. Theory is similarly not available for interval estimates or statistical testing.

V. THE STATE OF THE PROBLEM

This brief survey should show that there are vast lacunae in the statistical theory of estimation and testing of phylogenies from nucleotide sequences. The stranglehold which parsimony methods have on molecular evolutionary practice has served to inhibit investigation of statistical methods. This is less the fault of the advocates of parsimony than of the extreme empirical bias of molecular biologists, who have little patience for discussion of the methodology of data analysis.

That statisticians will be attracted to the problem in sufficient numbers to make any impact is in some doubt. The field is in that exciting early stage in which it is not yet recognized as an important problem, and therefore cannot attract funding. It seems likely that dubious methods will persist in this area for quite some time.

VI. ACKNOWLEDGMENTS

This work was supported in part by task agreement DE-AT06-76EV71005 of contract DE-AM06-76RL02225 between the U. S. Department of Energy and the University of Washington.

Chapter 7

CONVERGENT EVOLUTION AND NONPARAMETRIC INFERENCES FROM RESTRICTION DATA AND DNA SEQUENCES

Alan R. Templeton

Department of Biology, Washington University, St. Louis, Missouri

I. INTRODUCTION

Evolution in its most basic sense is a change in the genetic composition of a population through time. Because evolution is a genetic process, it is not surprising that recombinant DNA techniques can be applied to evolutionary studies. One major area of applicability is the reconstruction of phylogenies. In this chapter, I deal with the problem of making phylogenetic inference from restriction endonuclease site maps and DNA sequence data.

Restriction site maps are generated by isolating a segment of DNA and then subjecting the DNA to cleavage by a variety of restriction endonucleases--enzymes which cleave the DNA whenever a particular recognition sequence occurs. Different restriction enzymes have different recognition sequences, and hence each enzyme makes a direct, but different, assay of the DNA. In general, a battery of several different restriction enzymes is used on the DNA, and the various cleavage sites are mapped onto the DNA segments being studied.

Avise *et al*. (1979) have pointed out that there are two basic approaches for generating phylogenetic inferences from restriction maps. In the first type of analysis, each restriction site is regarded as a separate character, and common restriction patterns are assigned common evolutionary origins. Phylogenies are constructed by maximum parsimony.

In the second approach, some sort of distance measure is constructed between all species pairs. A phylogeny is then inferred from these distances, usually by some phenetic algorithm. A variety of distance measures applicable to restriction site data have been recently proposed (Gotoh *et al*., 1979; Kaplan and Langley, 1979; Nei and Li, 1979; Engels, 1981; Kaplan and Risko, 1981). The Nei and Li (1979) distance is one of the more popular measures, perhaps because of its simplicity. Also, the more complex distance measures give very similar results to those of Nei and Li (Kaplan and Risko, 1981). Consequently, I will focus on the Nei and Li distance measure in this chapter.

Unfortunately, there are several difficulties with both types of analyses. Both of them effectively ignore the problem of convergent evolution, *i.e.*, the sharing of a common restriction pattern in two present day species, *not* because the pattern was present in a common ancestor, but because it evolved independently in the two lineages. One of the primary purposes of this chapter is to show that convergent evolution should not be ignored.

A second difficulty of these analyses is that they have traditionally used data pooled across several different restriction enzymes. Indeed, such pooling is often explicitly recommended (Gotoh *et al*., 1979; Nei and Li, 1979). However, Adams and Rothman (1982) have recently analyzed the distribution of the frequencies and of the locations of recognition sites of 54 restriction endonucleases for several pieces of DNA that have been completely sequenced. Their results show that there is substantial nonrandomness in the distribution of restriction sites in all DNA sequences examined, and that considerable heterogeneity exists between different restriction enzymes. Moreover, the assumptions concerning DNA evolution that underlie the Nei and Li distance, and the simulations used to justify this measure (Nei and Li, 1979; Kaplan and Risko, 1981; Li, 1981) were found not to be good approximations of reality. As a consequence of all these factors, Adams and Rothman concluded that distance measures would be biased and would be unreliable indicators of phylogenetic relationships.

A third difficulty with phenetic analysis is that many phenetic algorithms require, or at least work best under, the assumption that the rate of evolution is constant over time and is identical in all lineages (Farris, 1981). However, this molecular clock hypothesis has recently been questioned by several authors (Corruccini *et al.*, 1980; Jacobs and Pilbeam, 1980; Naylor, 1980; Goodman, 1981) and has been specifically rejected for restriction site evolution in primate mitochondrial DNA (Templeton, 1983).

A final difficulty with almost all algorithms for generating phylogenies is that they can identify the "best" phylogeny by some criterion, but rarely do they allow an evaluation of the statistical significance of that statement. For example, is the best phylogeny statistically better than the second best (see also the discussion in the chapter by Felsenstein in this volume)?

The purpose of this chapter is first to discuss the problem of convergent evolution, and then to outline an algorithm that deals with the difficulties mentioned above. A detailed worked example of the use of this algorithm is given in Templeton (1983) and will not be presented here. However, some of the results of that analysis will be presented here when they are relevant to the general problems being considered. Finally, an extension of the restriction site testing procedures is given for testing alternative phylogenies from DNA sequence data.

II. THE PROBLEM OF CONVERGENT EVOLUTION

The basic idea underlying all the distance measures is that shared traits from the ancestral state are gradually lost with time and are replaced by new unshared traits. For example, the Nei and Li distance measure is based upon S, the proportion of shared restriction sites between two lines. Using Nei and Li's terminology, let n_x be the number of restriction sites in lineage x, n_y the number in lineage y, and n_{xy} the number shared by x and y. Then

$$S = 2n_{xy}/(n_x + n_y)$$

and the genetic distance is simply -log(S)/r where r is the number of nucleotides in the recognition sequence. Although S is the observed proportion of shared sites, Nei and Li (1979) describe it as "the proportion of ancestral restriction sites that remain unchanged in both lines". If convergent evolution does occur, it means that the observed S does not correspond to the Nei and Li definition of S. Since the distance measure is based upon observed similarity, the evolutionary interpretation of their distance becomes unclear if convergent evolution is important.

I will consider four types of convergent evolutionary events, as shown in Figure 1. The first type of convergence occurs when a mutation destroys an ancestral restriction site in one lineage, but a second mutation restores the site at a later time. Thus, two lineages can share an ancestral site even though that ancestral site is not "unchanged". Moreover, as illustrated in Figure 1a, if speciation occurred between the first mutational event and the second, the two species most closely related would not share this site (unless a second restoring mutation occurred - a very unlikely event), whereas two of the more distantly related species would share this site. I will call this type of convergence a loss-gain. Nei and Li (1979) calculated the impact of this type of convergent evolution upon their distance measure and concluded that its impact would be minimal "unless λt is larger than about 0.15", where the rate of nucleotide substitution is λ and t is the time since the two lines split. This is the only type of convergence explicitly considered by Nei and Li (1979), and it will be seen that it requires two mutational events. However, there are three other types of convergent evolution that also require two mutational events (Figure 1). Additional types of convergent events are possible, but they will involve three or more mutational events and will not be considered.

Figure 1b shows a type of convergence in which the two lines share a restriction site not found in the ancestor. To understand

a. LOSS-GAIN b. CONVERGENT GAIN

ANCESTOR: SITE -1 SITE

CURRENT SPECIES: SHARED SITE NO SITE SHARED NON-ANCESTRAL SITE NO SITE

c. CONVERGENT LOSS d. GAIN-LOSS

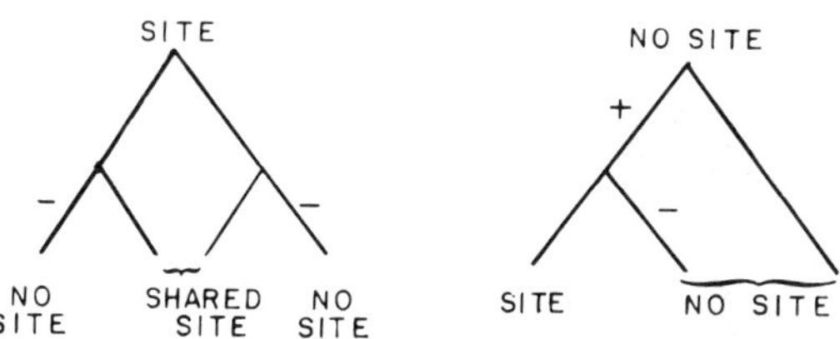

Figure 1. Four basic types of convergent evolution involving restriction sites. (a) Ancestral site is lost in one lineage but regained due to a subsequent mutation. (b) One-off site in the ancestral DNA mutates to restriction site in two independent lineages, causing the two species to share a site not present in the ancestor. (c) Ancestral site lost in two independent lineages. (d) Site gained after speciation and subsequently lost due to second mutational event. All four of these types of events can lead to a situation in which two distantly related species share a pattern not shared by closely related species.

how this could occur, it is important to note that in the ancestral state the entire DNA sequence is shared and not merely that subset recognized by a particular restriction enzyme. As time proceeds, the two lines will acquire new restriction sites, primarily from those DNA segments in the ancestor that already have the recognition sequence apart from one nucleotide. I will call such nucleotide sequences one-off sites (symbolized by "-1" in Figure 1). Because the lineages share not only ancestral restriction sites, but also ancestral one-off sites, the new sites that first appear in the two lineages come primarily from a small subset of the DNA that is the same in both lines. If, by chance, the same one-off sites are mutated into restriction sites in two lineages, a convergent gain will occur. Li (1981) calculated the probability of convergent gains and concluded that there is no need to take them into account as long as λt is less than or equal to 0.15.

The third type of convergence is convergent loss, as illustrated in Figure 1c. Under convergent loss, the same ancestral restriction site is lost in two independent lineages. As portrayed in Figure 1c, convergent loss can also result in more distantly related species sharing a site that is not shared between more closely related species. Hence, this type of convergence should not be ignored when evaluating the validity of distance measures or cladistic algorithms.

The final type of convergence, gain-loss, is shown in Figure 1d. In this case, a site is gained through mutation in a lineage and then subsequently lost. Once again, as illustrated in Figure 1d, this type of event can also obscure phylogenetic relationships.

I will now quantify the importance of these types of convergence. To make my results comparable to those of Nei and Li (1979) and Li (1981), I will assume a model of DNA evolution virtually identical to that of those authors; namely, that the DNA initially consists of a random sequence of nucleotides, that mutations occur at random and uniformly over the DNA molecule with a constant rate of λ per unit time, and that the appropriate sampling model of nucleotide substitutions is a Poisson process. However, I will examine the importance of violating one of their assumptions: that a nucleotide mutates to any of its three alternatives with equal probabilities. Recent work by Brown *et al.* (1982) strongly indicates that transitions are far more likely than transversions in their primate mitochondrial DNA data (90% of the mutations are transitions, whereas the equal probability assumption predicts that only 1/3 of all mutations should be transitions). (See also the evidence given by Brown and Clegg in their chapter in this volume.)

As mentioned earlier, Nei and Li (1979) considered the impact of loss-gain convergences upon their distance measure. They examined this type of convergence by letting p_t be the probability that a nucleotide at a particular site at time t is the same as that at time zero. Then, with the assumptions about mutation given earlier, they derived the recurrence formula for p_t as

$$p_{t+1} = (1 - \lambda)p_t + \lambda(1 - p_t)/3 \tag{1}$$

They approximated the solution to (1) by a continuous time solution with $p_0 = 1$ and obtained

$$p_t = (1 + 3e^{-4\lambda t/3})/4 \tag{2}$$

For a restriction site to exist at the original position, all of the nucleotides in the recognition sequence must be identical with the original ones (actually, for some restriction enzymes the recognition sequence allows more than one nucleotide at some sites, but I will ignore this problem, as did Nei and Li). Let r be the number of nucleotides in the recognition sequence. Then, the probability that a restriction site exists at time t that was present at time zero is $P = (p_t)^r$. Nei and Li also showed that, under these assumptions, the probability of an original restriction site being unchanged by mutation up to time t is given by $\exp(-r\lambda t)$.

Since I am considering only convergent events that involve at most two mutations, a site present at time t under this approximation exists either because it was unchanged by mutation or because a loss-gain occurred. Moreover, these are mutually exclusive events. Hence, the probability of a loss-gain at time t is given by

$$(p_t)^r - e^{-r\lambda t} \tag{3}$$

Table 1 presents some values of (3) as a function of λt for $r = 6$, a common value for many restriction enzymes. In addition, this table gives the values for $\exp(-r\lambda t)$, the probability of sites remaining unchanged.

Table 1. Probability of an ancestral restriction site remaining unchanged and probability of a loss-gain occurring with r=6. The probability of a mutation being a transition is u.

	Probability	Probability of a Gain-Loss	
λt	Site Unchanged	u = 1/3	u = 0.9
0.0	1.0000	0.0000	0.0000
0.01	0.9418	0.0001	0.0002
0.02	0.8869	0.0004	0.0009
0.03	0.8353	0.0008	0.0018
0.04	0.7866	0.0013	0.0031
0.05	0.7408	0.0019	0.0046
0.075	0.6376	0.0037	0.0089
0.100	0.5400	0.0056	0.0137
0.125	0.4724	0.0076	0.0185
0.150	0.4066	0.0096	0.0231
0.200	0.3012	0.0128	0.0312

The above calculations are all based upon the assumption that, when a nucleotide mutates, it is equally likely to mutate to any of its three alternatives. However, in view of the work of Brown *et al.* (1982), this is not a realistic assumption. Consequently, I will now consider a more generalized model in which u is the probability of the mutation being a transition, and $v = 1 - u$ is the probability that it is a transversion. The assumption that all nucleotides are equally likely is the special case $u = 1/3$. With this model of mutation, the analog of equation (1) is

$$p_{t+1} = (1 - \lambda)p_t + \lambda u^2(1 - p_t) + \lambda v^2(1 - p_t)/2 \tag{4}$$

Note that (4) reduces to (1) when $u = 1/3$. The continuous time solution to (4) when $p_0 = 1$ is

$$p_t = \{u^2 + v^2/2 + \exp[-\lambda(1 + u^2 + v^2/2)t]\}/(1 + u^2 + v^2/2) \tag{5}$$

As expected, this reduces to (2) when $u = 1/3$. The probability of a loss-gain can now be calculated from (3) using (5) to evaluate p_t. Some numerical examples of the probability of a loss-gain under this model are given in Table 1 for $r = 6$ and $u = 0.9$ (the value suggested by the data of Brown *et al.*, 1982). As can be seen, the probability of a loss-gain goes up considerably, being about 2 to 2.5 times more likely than when $u = 1/3$.

The second type of convergence in Figure 1 is convergent gain. Li (1981) calculated the impact of this type of convergence as follows. First, under the assumptions of equal base content and random base distribution, the expected number of one-off sites in the ancestor for a DNA segment of m bases is $mr(3/4)(1/4)^{r-1} = 3rma$, where $a = (1/4)^r$. At time t, the probability for the r-1 correct bases remaining unchanged is $\exp(-\lambda t)^{r-1}$ and the probability of the wrong base changing to the correct base is $[1 - \exp(-\lambda t)]/3$. Thus, the probability of a one-off site becoming a site at time t is

$$q = e^{-(r-1)\lambda t}(1 - e^{-\lambda t})/3 \tag{6}$$

and the probability of this occurring in both lineages is q^2. However, to make the probability q^2 comparable to the probabilities given above for loss-gains, it is important to note that for every ancestral site, there are 3r one-off sites under the assumptions made by Li (1981). Consequently, q^2 should be multiplied by 3r to measure the frequency of occurrence of convergent gains on a per ancestral site basis. Numerical values for $3rq^2$ are given in Table 2 for r = 6.

Table 2. Frequency of convergent gains with r=6.

λt	Li's Equation	Equation (16)
0.00	0.0000	0.0000
0.01	0.0002	0.0002
0.02	0.0006	0.0006
0.03	0.0013	0.0013
0.04	0.0021	0.0020
0.05	0.0029	0.0028
0.075	0.0049	0.0048
0.100	0.0067	0.0065
0.125	0.0079	0.0077
0.150	0.0087	0.0084
0.200	0.0089	0.0085

By contrasting the appropriate columns of Tables 1 and 2, it is evident that convergent gains are similar in frequency of occurrence to loss-gains under the assumptions of the Nei and Li model, although convergent gains are slightly more likely for all values of λt tabulated. This rough equality in frequency is not surprising. For a loss-gain, the first mutation can occur in any one of 3r ways and the second mutation must exactly reverse the first. For a convergent gain to occur, both mutations must be exactly the same, but there are 3r more one-off sites than sites. Consequently, on a per ancestral site basis, there are 3r ways of achieving a loss-gain through two mutations, and 3r ways of achieving a convergent gain through two mutations.

I will now examine the effect of $u \neq 1/3$ on the probability of convergent gains. Once again, I will ignore all events requiring more than two mutations. Under this assumption, the probability of r-1 correct bases remaining unchanged by time t is, as given by Li, $[\exp(-\lambda t)]^{r-1}$. However, I will divide the one-off bases into two categories: those that differ from a recognition sequence by one transition, and those that differ by one transversion. With a probability of $u[1 - \exp(-\lambda t)]$ a base in the first category mutates to the correct one, while the probability for second category bases is $v[1 - \exp(-\lambda t)]/2$. Let w be the proportion of one-off sites that differ from a recognition sequence by a transition, and 1-w the proportion differing by a transversion. Then, the total probability of a one-off site becoming a site at time t is

$$q = e^{-(r-1)\lambda t} (1 - e^{-\lambda t}) [uw + v(1 - w)/2] \qquad (7)$$

The Nei and Li model is basically one of random base sequences in an equilibrium state. The equilibrium state has been shown by Adams and Rothman (1982) not to depend upon u, and always has one-off sites differing by a transversion being twice as common as one-off sites differing by a transition. Hence, at equilibrium, $w = 1/3$, and equation (7) reduces to (6) and does not depend at all on u or v. Consequently, to the limits of the approximations used here (especially the ignoring of events involving more than two mutations), the probability of transitions versus transversions does not have any impact on the probability of this type of convergence. This means, however, that the relative probabilities of loss-gains versus convergent gains are functions of u. For $u = 1/3$, convergent-gains are slightly more probable than loss-gains, while, for $u = 0.9$, this order is reversed (as can be seen from Tables 1 and 2). For all values of u in this range, the probabilities of the two convergence types are of the same order of magnitude.

It is now possible to measure the joint effect of increased loss-gains (due to high u values) and convergent losses on the Nei and Li distance measure. By adding the probability of loss-gains in the $u = 0.9$ column of Table 1 to the appropriate column of Table 2, the same percentage deviation from the linearity claimed at $\lambda t = 0.15$ by Nei and Li (1979) now occurs at $\lambda t = 0.08$. This joint consideration of events thus suggests that the Nei and Li distance measure breaks down twice as fast as they found under their assumptions.

Before going on to other types of convergence, I will consider an alternative derivation of the frequency of convergent gains that will be of use later. It will be sufficient to consider the case only of $u = 1/3$. Adams and Rothman (1982) give a general set of differential equations for the dynamics of changes in numbers of sites, one-off sites, two-off sites and so on. Considering only the case where $u = 1/3$, so that there is no need to distinguish between one-off sites differing by a transition and those differing by a transversion, and ignoring any event that can yield convergence with three or more mutations, their equations simplify to

$$\begin{aligned} \frac{de_0}{dt} &= \lambda(e_{-1}/3 - re_0) \\ \frac{de_{-1}}{dt} &= \lambda[re_0 - (r - 2/3)e_{-1}] \end{aligned} \tag{8}$$

where e_0 is the number of restriction sites and e_{-1} is the number of one-off sites. Letting $E(t)$ be the column vector of e_0 and e_{-1} at time t from divergence and

$$A = \lambda \begin{bmatrix} -r & 1/3 \\ r & -r+2/3 \end{bmatrix} \tag{9}$$

equation (8) can be rewritten as

$$\frac{d\,E(t)}{dt} = A\,E(t) \tag{10}$$

The solution to (10) is

$$E(t) = e^{At}\,E(0) \tag{11}$$

where e^{At} denotes the matrix exponential of At. The number of new sites gained in a lineage by time t from the original one-off category is now easily determined by calculating the number of sites $e_0(t)$ at time t under the initial conditions

$$E(0) = \begin{bmatrix} 0 \\ \\ e_{-1}(0) \end{bmatrix} \tag{12}$$

The frequency of convergent gains can now be found as follows. Let $e_0(t)$ be the number of new restriction sites in one lineage that have evolved from the one-off category since the lines split t time units ago, and $e'_0(t)$ the corresponding number in the second lineage. Then, from elementary sampling theory, the probability distribution function for z(t), the number of matches (*i.e.*, convergent gains) derived from mutations in the ancestral one-off sites, is

$$\frac{\begin{pmatrix} e_0(t) \\ \\ z(t) \end{pmatrix}\begin{pmatrix} e_{-1}(0) - e_0(t) \\ \\ e'_{(0)}(t) - z(t) \end{pmatrix}}{\begin{pmatrix} e_{-1}(0) \\ \\ e'_0(t) \end{pmatrix}} \tag{13}$$

where large parentheses denote the combinatorial operator. From (13), the expected value of z(t) is

$$e_0(t)\ e'_0(t)\ /\ e_{-1}(0) \tag{14}$$

Equation (14) is a conditional expectation given $e_0(t)$ and $e'_0(t)$. In general, these e's should be treated as random variables. Moreover, assuming that both lineages are identical with respect to the underlying forces affecting restriction site evolution, both e's should be independent and identically distributed. Hence, the unconditional expectation of z(t) is approximately

$$E[e_0(t)]^2 \ / \ e_{-1}(0) \tag{15}$$

The expected value of $e_0(t)$ is given by equation (11) under the initial conditions (12), so equation (15) becomes

$$E[z(t)] = \frac{e_{-1}(0)\ e^{[-2\lambda t(r-1/3)]}\ (B + 1/B - 2)}{(12r + 4)} \tag{16}$$

where

$$B = e^{\{2\lambda t\ [(r + 1/3)/3]^{\frac{1}{2}}\}}$$

Once again, to make this comparable to the probability of loss-gains on a per ancestral site basis, simply substitute 3r for $e_{-1}(0)$ in equation (16). A numerical illustration of the results is given in Table 2, and, as can be seen, the results obtained with this approach are nearly identical to those obtained by Li (1981).

I will now consider convergent losses. Once the probability of an original site still being present at time t is known, it is a simple matter to calculate the probability of a convergent loss. This probability is simply

$$(1 - p_t^r)^2 \tag{17}$$

where either (2) or (5) can be used to evaluate p_t, depending on which model of mutation is desired. Table 3 presents some numerical evaluations of (6) under both models. As can be seen from this table, the probability of this type of convergent evolution is decreased when transitions are more probable than under the random mutation model. However, as also is evident from Table 3, this decrease is very small even when u = 0.9 and when λt is up to 0.2. Consequently, having an increased frequency of transitions greatly increases the probability of loss-gains but minimally decreases the probability of convergent loss.

Table 3. Frequency of convergent loss (r=6).

λt	u=1/3	u=0.9
0.00	0.0000	0.0000
0.01	0.0034	0.0034
0.02	0.0127	0.0126
0.03	0.0269	0.0265
0.04	0.0450	0.0442
0.05	0.0662	0.0648
0.075	0.1287	0.1250
0.100	0.1985	0.1914
0.125	0.2704	0.2592
0.150	0.3409	0.3252
0.200	0.4706	0.4457

By contrasting Table 3 with Tables 1 and 2, it is obvious that convergent loss is more probable, by an order of magnitude, than loss-gains or convergent gains under either mutation model. This result is expected. On a per ancestral site basis, as above, there are 3r ways of achieving loss-gain and convergent gain with two mutations. However, any one of 3r mutations in one lineage coupled with any one of 3r mutations in the second lineage will result in a convergent loss, so that this event can occur in $(3r)^2$ ways. The probabilities of convergent losses are thus an order of magnitude greater than those for loss- or convergent gains for r of four to six.

The final type of convergence, gain-loss, should be of the same order of magnitude as convergent loss. The gains come from one-off sites and so need a specific mutation to occur. Once they have become a site, however, any one of 3r mutations will cause a loss. Once again, to adjust this to a per ancestral site basis, it must be recalled that there are 3r one-off sites for every site, and hence the number of ways of generating a gain-loss from two mutations is also $(3r)^2$. One might expect loss-gains to be somewhat less probable than convergent losses because most mutations in the one-off sites cause them to become two-off sites, and hence to be ignored at the level of approximation used here. This causes the number of ways of generating gain-losses to rapidly decrease with time, compared to convergent losses.

I return to equation (11) to quantify the frequency of gain-losses, and solve for $e_{-1}(t)$ with initial conditions (12). Under these initial conditions, the one-off sites at time t are of only two types: those that have remained one-off sites, and those that have mutated into a restriction site and then mutated back to a one-off site (*i.e.*, gain-losses). With the present approximations, there is no input to the one-off category from the two-off category, and one-off sites that mutate into two-off sites are lost from the process. Hence, the frequency of gain-losses is simply the frequency of one-off sites that are present at a given time minus the frequency of one-off sites that remained so throughout. To remain a one-off site, there must be no mutations at the r-1 correct bases, and mutation only to another incorrect base is allowed at the incorrect base. With u = 1/3, the probability of this joint event is

$$e^{-r\lambda t} + 2e^{[-(r-1)\lambda t]}(1 - e^{-\lambda t}) / 3 \tag{18}$$

Hence, using equations (11), (12) and (18), and adjusting for there being 3r one-off sites for every initial site, the frequency of gain-losses per original restriction site is given by

$$(3/2)r\ e^{-(r-1/3)\lambda t}\ [B(1+C) + (1-C)/B]$$
$$- 3r[e^{-r\lambda t} + 2e^{-(r-1)\lambda t}\ (1 - e^{-\lambda t})] \tag{19}$$

where

$$B = e^{\lambda t[(r+1/3)/3]^{\frac{1}{2}}},\quad C = 1 / (3r + 1)^{\frac{1}{2}}$$

A numerical evaluation of equation (19) is given in Table 4 for r = 6. The frequency of gain-losses is seen to be less than the frequency of convergent losses, although the frequencies of both convergent losses and of gain-losses are of the same order of magnitude, and both about an order of magnitude greater than those for loss-gains or convergent gains. This confirms earlier predictions.

Table 4. Frequency of gain-losses (with r=6).

λt	Frequency
0.00	0.0000
0.01	0.0015
0.02	0.0057
0.03	0.0122
0.04	0.0205
0.05	0.0303
0.075	0.0593
0.100	0.0919
0.125	0.1250
0.150	0.1568
0.2000	0.2118

Given that convergent losses and gain-losses are by far the most common types of convergence under the present assumptions, what effects do they have on the Nei and Li distance? This distance, supposed to measure "the mean number of nucleotide substitutions per nucleotide site", has an ideal value of $2\lambda t$. The bias in the distance from ignoring convergent losses and gain-losses can be quite serious since actual distances can be underestimated by two mutations for each convergent event that occurs. For $\lambda t = 0.03$, Tables 3 and 4 indicate that about as many mutations are ignored as are scored in the Nei and Li distance.

To illustrate the importance of convergence, I have analyzed the data set on human and ape mitochondrial DNA given by Ferris *et al.* (1981). The rate of mitochondrial nucleotide substitution has been estimated as an average of 0.01 per million years from this data (Ferris, personal communication). The oldest divergence in this human-ape phylogeny is the gibbon divergence, which occurred about 12 million years ago (Andrews, 1982). Thus, λt is less than about 0.12 in the entire data set, well within the range where convergence was thought safe to ignore (Nei and Li, 1979).

Table 5 gives the Nei and Li distance [calculated from equation (13) of Nei and Li (1979)] obtained from data on 18 restriction enzymes with $r = 6$. The data set had been generated in the hopes of resolving the evolutionary relationships between man, chimps and gorillas. As can be seen from Table 5, however, the distances between these three species are quite similar and are statistically indistinguishable when their standard deviations are considered. The phylogenetic information in the data set is not revealed by the distance measure, although it can be extracted by a nonparametric procedure (Templeton, 1983).

Table 5. Nei and Li genetic distances calculated from mitochondrial DNA (18 restriction and endonucleases with r=6, Ferris *et al.*, 1981).

Species Compared	Distance	Standard Deviation
Gorilla-Chimp	0.084	0.019
Human-Gorilla	0.102	0.022
Human-Chimp	0.125	0.025
Human-Orang	0.105	0.022
Gorilla-Orang	0.143	0.028
Chimp-Orang	0.153	0.029
Human-Gibbon	0.153	0.029
Gorilla-Gibbon	0.145	0.028
Chimp-Gibbon	0.174	0.033
Orang-Gibbon	0.174	0.033

As part of the nonparametric analysis (Templeton, 1983), I reconstructed the mutational events responsible for the data set under a criterion to be discussed in the next section for the evolutionary tree shown in Figure 2. The results are given in Table 6, where 30% of all mutations in the tree are seen to be involved in convergent events. This 30% was constructed to be a *minimum* estimate, so that, even for $\lambda t < 0.15$, convergent events should not be considered unimportant. Further, most convergent events are the convergent losses or gain-losses ignored by Nei and Li.

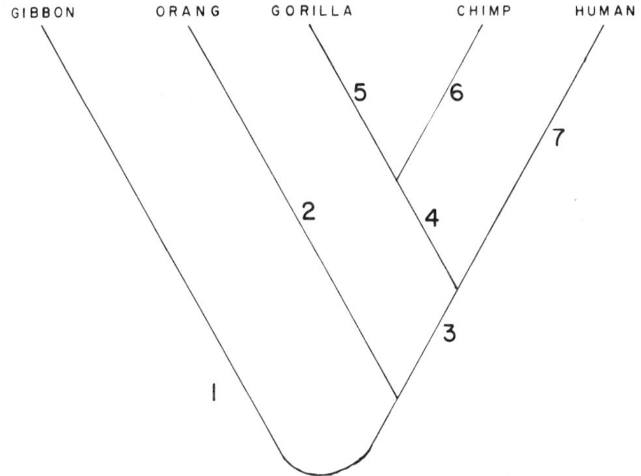

Figure 2. Segments of the human-ape phylogeny. Each portion of the phylogeny between two branching points or a branching point of an existing species is given a number (1-7). The bottom of the phylogeny is drawn in a rounded fashion to indicate that the exact ancestor of the five species cannot be specified from this data.

III. A NONPARAMETRIC ANALYSIS OF RESTRICTION SITE DATA

The previous section shows that a genetic distance approach is not satisfactory for making phylogenetic inferences from restriction site data. To avoid the difficulties mentioned in the previous section and in the Introduction, I have devised algorithms for estimating and testing phylogenies (Templeton, 1983). Here I will discuss how these algorithms deal with the difficulties and accommodate convergent evolution.

A. Phylogeny Estimation

Unlike most estimation algorithms for restriction sites, my procedure does not pool information obtained from several different restriction enzymes. Rather, the data sets generated by a particular restriction enzyme are treated as the fundamental units of analysis. The algorithm involves two steps.

Table 6. Convergent events for phylogeny of Figure 2.

Enzyme	Site	Type of Convergence	Location*
f. *Bam* HI	1	Gain-loss or Convergent loss	1-5
	2	Convergent loss	2-7
h. *Pvu* II	5	Convergent loss	2-7
	6	Convergent loss	2-6
i. *Sal* I	1	Convergent gain	2-5
l. *Xho* I	1	Gain-loss or Convergent loss	1-4
m. *Ava* I	1	Convergent gain	2-6
	2	Convergent loss	2-4
o. *Hinc* II	2	Gain-loss	3-6
	3	Gain-loss or Convergent loss	1-4
	10	Gain-loss or Convergent loss	1-6
	19	Loss-Gain or Convergent gain	1-6
w. *Bst* EII	1	Loss-gain or Convergent gain	1-6
	2	Convergent loss	1-5
	4	Convergent loss	2-4
x. *Bcl* I	6	Gain-loss	3-5
	7	Gain-loss	3-6
	10	Loss-gain or Convergent gain	1-6
y. *Bgl* I	5	Gain-loss or Convergent loss	1-5
z. *Fnu* DII	4	Gain-loss	3-5
	5	Convergent loss	2-5

*The location of the two mutational events causing convergence are indicated in terms of the numbered segments given in Figure 2.

The first step is to generate, through maximum parsimony, an unrooted tree for each enzyme-specific data set. If a root can be determined by outside data, these trees are additionally constrained to minimize the number of convergent gains or loss-gains relative to the number of convergent losses or gain-losses. This constraint is justified by the calculations of the previous section that show convergent gains and/or loss-gains are about an order of magnitude less probable than convergent losses or gain-losses over a very broad range of λt values.

Convergence can cause many of the enzyme-specific trees constructed in step 1 to be inconsistent with the true phylogeny. These inconsistent trees will be incompatible with any trees consistent with the true phylogeny and often with each other. One way of dealing with incompatible data sets is to perform a compatibility analysis (Estabrook *et al.*, 1976).

The second step of the estimation algorithm is to perform such an analysis between the enzyme-specific trees constructed in step 1. This analysis estimates the true phylogeny as the phylogeny with maximal topological consistency; that is, the tree that is consistent with more enzyme-specific trees than is any alternative tree. It is important to note that, although maximum parsimony was used in the first step of estimation, the tree resulting from the compatibility analysis is not necessarily a maximum parsimony tree.

Felsenstein (1981b) has shown that compatibility is more consistent than parsimony when some characters have a high rate of evolution and others do not. Such rate heterogeneity is expected between restriction enzyme data sets (Adams and Rothman, 1982), so compatibility should be superior to parsimony in analyzing such data. This superiority stems from some well-known limitations of the maximum parsimony procedure. Maximum parsimony tends to be consistent when the probability of a restriction site being altered by multiple mutational events (*i.e.*, convergence) is much smaller than the probability of a site being mutated only once, and when the rates of mutational substitution are about equal in all lineages; otherwise, inconsistency can occur (Cavender, 1978; Felsenstein, 1978b, 1979, this volume).

As shown in this chapter, convergent evolution of restriction sites is not very likely when λt is less than 0.05, but the probability of convergence increases rapidly after that point. The values of t are determined by history, but the values of λ will vary greatly from one restriction enzyme to another (Adams and Rothman, 1982). Consequently, even if the average λt was of the order of 0.05 for a particular data set, some restriction enzymes would have λt values well above 0.05 and others below that value. This means that some enzyme data sets would have little convergent evolution, and hence be in a parameter domain for which parsimony tends to be consistent (except for certain deviations in rate constancy, as discussed shortly), whereas others would have much convergent evolution and hence be in a parameter domain where parsimony tends to be inconsistent. Performing maximum parsimony upon the pooled data set is therefore tantamount to mixing good data with bad. However, compatibility tends to eliminate those enzyme-specific data sets with the higher λt values that are most affected by convergence (Templeton, 1983). Instead, the phylogenetic inference is drawn primarily from those data sets least affected by convergence, and hence those that are most likely consistent. By making inference from data sets with slow rates of evolution, the problem of inconsistency due to unequal rates of evolution is also minimized. As shown by Felsenstein (this volume), the evolutionary rates have to become increasingly disparate for parsimony to become inconsistent as the average evolutionary rate becomes smaller. Hence, compatibility should be robust to all but extreme deviations from rate constancy when λt values are around 0.05.

Note that the compatibility analysis effectively discards some of the data in making phylogenetic inference. Perhaps then, the consistency is achieved at the price of lower efficiency. However, as convergent events accumulate with appreciable frequency, phylogenetic information is destroyed. Compatibility tends to throw out the least informative data sets, so its efficiency is quite good (Penny, 1982a). In summary then, this estimation algorithm is both efficient and consistent as long as average λt values are 0.05 or less, and is robust to all but extreme deviations from the molecular clock hypothesis.

However, if the average λt values are well above 0.05, this estimation procedure is poor and should not be used. This may or may not be a serious problem, depending on the type of statistical inference desired. If one wishes to reconstruct the evolution of a specific DNA segment, this problem is serious for large λt values, and some other algorithm should be used. On the other hand, if one wishes to reconstruct the evolutionary relationships of a group of species and DNA is only being used as a tool for making that reconstruction, then this is not a serious problem. One of the more exciting discoveries emerging from the use of recombinant DNA techniques is that different pieces of DNA are evolving at greatly different rates. Hence, if one wishes to make phylogenetic inferences about a particular group of species, one simply finds a piece of DNA such that average λt values are around 0.05 or less. This is illustrated in Templeton (1983) where relationships between humans, chimpanzees and gorillas could be made from mitochondrial DNA, but the relationship between humans and orangutans could not because of too much convergent evolution. The relationship between humans and orangutans was easily estimated from globin DNA, however, as this had a slower rate of evolution than did mitochondrial DNA.

B. Testing Phylogenetic Hypotheses

Given that one has estimated a phylogeny by the above algorithm, or by some other procedure, it is often desired to test the significance of the difference between that phylogeny and some alternative. Unfortunately, few procedures exist for such testing, although one of the few is the maximum likelihood method discussed by Felsenstein in this volume. Maximum likelihood is often difficult computationally and is a parametric procedure. Felsenstein (1983) has also pointed out that maximum likelihood techniques are sometimes sensitive to even small departures from the assumptions used in generating the underlying parametric model. Given our ignorance about the biological appropriateness of the assumptions and about precisely how robust the likelihood procedure is to deviations from these assumptions, nonparametric procedures seem to be preferable to parametric procedures when an option exists, as it does for restriction site data (Templeton, 1983).

The first step in the nonparametric testing procedure is to construct the maximum parsimony/minimum convergent gain and/or loss-gain trees for each enzyme data set under the assumption that the estimated phylogeny (phylogeny A) is true, and then under the assumption that the alternative (phylogeny B) is true. A matched scoring procedure is then invoked to contrast how well a specific enzyme tree does under the A versus B comparison. Whenever a single mutational event in A is replaced by a convergent loss or gain-loss in B, a score of 1 is assigned. If, on the other hand, B had the single event and A the convergent loss or gain-loss, a score of -1 is given. Whenever a convergent loss or gain-loss in A is replaced by a convergent gain or loss-gain in B, a score of 1 is again assigned (with a -1 for the opposite replacement). If a single mutation in A is replaced by a convergent gain or loss-gain in B, a score of 2 is assigned (with a -2 for the opposite replacement). These scores are then summed over all events in the enzyme specific trees and a matched tree score is obtained.

The scoring procedure results in a rank ordering by two criteria. First, single events are deemed more likely than convergent events, and, as mentioned earlier, this is a good assumption when average λt values are of the order of 0.05 or less. Although this assumption can be violated, the nonparametric test loses statistical power when convergence is common (Templeton, 1983) and λt values are large. Hence, whenever an alternative hypothesis can be rejected, it is because this ranking criterion holds true and the test effectively contains a self-destruct mechanism that prevents statistical inference being obtained when its underlying assumptions are violated. This is a strength, not a weakness, of the procedure.

The second ranking criterion is simply that convergent gains and/or loss-gains are deemed less likely than convergent losses and/or gain-losses. As is evident from the equations given in this chapter, convergent gains and/or loss-gains are an order of magnitude less likely over the entire range of λt values for which this test has power. Although this conclusion rests upon the parametric model developed in this chapter, the ranking criterion should be robust to deviations from this model. As noted earlier, this inequality in relative probabilities stems from the fact that there are many more ways of achieving a convergent loss or gain-loss than there are of achieving a convergent gain or loss-gain. Unless mutational events are extremely clustered at specific nucleotides, it would be very difficult to construct a model in which convergent gains and loss-gains would be more frequent than convergent losses and gain-losses. Consequently, this ranking criterion should be extremely robust.

Once the matched tree scores have been calculated, a one-tailed Wilcoxon matched pair signed rank test (Seigel, 1956) is performed on the signed ranks of the paired tree scores. Standard statistical tables can then be used to make statistical inferences. Templeton (1983) gives worked examples of this test procedure. The test is computationally simple; it is nonparametric and therefore makes no detailed assumptions about the sampling properties of restriction sites affected by millions of years of evolution. It is based upon a robust ranking criterion when λt values are of the order of 0.05 or less, and it loses power then λt values exceed that order. These are all highly desirable properties.

IV. INFERENCE FROM DNA SEQUENCE DATA

Because of the desirable properties of this test, it would be useful to construct an analog appropriate for DNA sequence data. In restriction site analysis, the fundamental unit was the set of sites identified by a particular restriction enzyme. This is a natural unit, both experimentally and statistically. For DNA sequences, the natural unit would seem to be the individual nucleotide base pairs. However, it is already evident that heterogeneities exist in DNA sequences with respect to substitution rates, and in many cases information is available to indicate more homogeneous subsets (*e.g.*, coding versus noncoding regions, third codon sites versus first codon sites). Consequently, given enough knowledge about the DNA segment being analyzed, it may prove that grouping certain sets of nucleotides is the best procedure. As this is presently an unsolved problem, I will make the simple assumption that each individual nucleotide is the unit of analysis.

The Wilcoxon test can easily be applied to such nucleotide data. For each variable nucleotide site found in the species sequenced, a maximum parsimony tree is constructed assuming either phylogeny A , or its alternative B , is true. (If desired, phylogeny A can be estimated using the same estimation algorithm summarized in the previous section; that is, maximum parsimony on each individual variable nucleotide site followed by compatibility between sites). The scoring procedure can then be based simply on the number of mutational events required under each tree, but perhaps additional information can be incorporated into the scoring procedure. For example, it is known that transitions are more likely than transversions for at least some DNA molecules. If that seems to be a reasonable assumption for the DNA segment being studied, then the paired tree score could be calculated by taking the number of transitions plus twice the number of transversions required under phylogeny B minus the corresponding quantity required under phylogeny A . Once again, this gives a ranking of the relative probability of events under the assumption of parsimony and of transitions being more likely than transversions. The Wilcoxon test can then be used on these scores.

As an example of this procedure, I have analyzed the DNA sequence data given in Brown *et al.* (1982) on 896 base pairs of the mitochondrial DNA of humans, chimpanzees, gorillas, orangutans and gibbons. I compared the six phylogenies given in Figure 3, regarding phylogeny 1 as the "best" phylogeny on the basis of the restriction site analysis (Templeton, 1983). The results of applying this nonparametric test are given in Table 7. Note that phylogenies 4, 5 and 6 are strongly rejected, but 2 and 3 are not. In contrast, when these same phylogenies were compared with the mitochondrial DNA restriction site data, numbers 2, 3 and 4 were rejected in favor of 1, but not 5 or 6. The reason for the failure

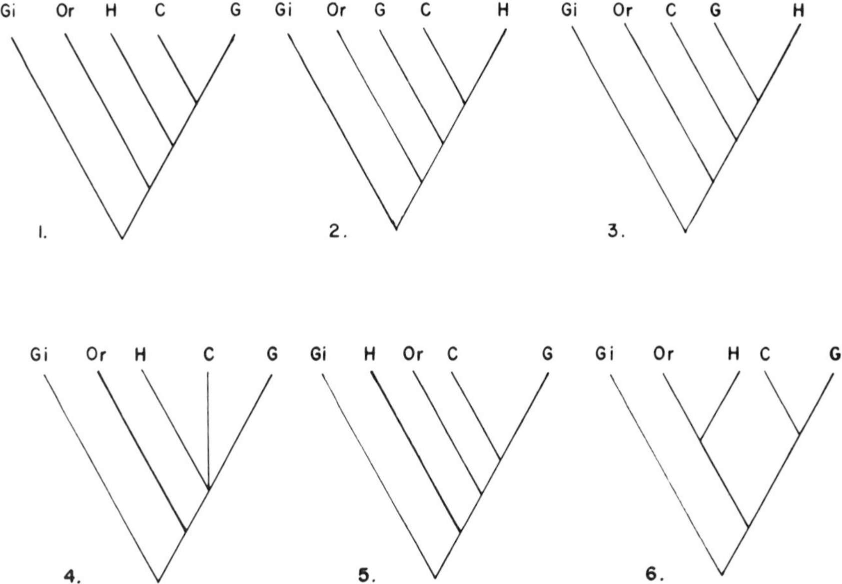

Figure 3. Some hypothetical primate phylogenies indicating the possible evolutionary relationships between gibbon (Gi), orangutan (Or), chimpanzee (C), gorilla (G) and human (H).

to discriminate between phylogenies 1, 2 and 3 with the sequence data is most likely a result of extensive polymorphism in the ancestor of human, gorilla and chimpanzee. Mitochondrial DNA is known to be highly polymorphic in these species when assayed by restriction enzymes (Brown, 1980), and the level of polymorphism should go up substantially when assayed by DNA sequencing. Templeton *et al.* (1982) showed that phylogenetic inference from genetic distances can be obscured by ancestral polymorphisms when dealing with recent speciation events, and the same is true for this type of analysis. Moreover, this problem becomes increasingly serious as levels of detectable polymorphism increase. DNA sequence data will therefore be more affected by this problem than will be restriction site data. Unfortunately, there is insufficient data on sequence polymorphism in these species to make any correction

Table 7. Wilcoxon matched-pairs signed ranks test of phylogeny 1 versus phylogenies 2 through 6 (Figure 3), based on DNA sequence data of 896 bp of mitochondrial DNA.

Phylogeny	Number of Comparisons	Score	Rank	Number of Nonzero Scores	Sum of Negative Ranks
2	12	+1	+10	22	-133
	7	-1	-10		
	3	-2	-21		
3	12	+1	+11	21	-99
	9	-1	-11		
4	12	+1	+6.5	12	0***
5	25	+1	+15.5	38	-77.5***
	5	-1	-15.5		
	8	+2	+34.5		
6	24	+1	+14	35	-42***
	3	-1	-14		
	8	+2	+31.5		

***Significant at 0.1% level.

for polymorphism at present. Hence DNA sequence data are, in general, less useful than restriction site data for reconstructing recent speciation events unless extensive population surveys are feasible. The problem of ancestral polymorphism diminishes with increased divergence times, however, as is evident from the strong discrimination involving the placement of humans and orangutans seen in Table 7.

Recall that the restriction data failed in the human-orangutan discrimination. The reason for this failure was simply that this contrast lay in a domain in which λt was too large (Templeton, 1983), and hence power was lost. Note, however, that this distinction is made quite easily from the DNA sequence data, as is to be expected. Convergent evolution at the nucleotide level can occur only when exactly the same site is mutated two or more times.

However, convergence for restriction sites occurs primarily by mutations at different nucleotide sites (convergent loss or gain-loss). Consequently, of the types of convergence considered in this chapter, convergent gains and loss-gains are the only types comparable to the types of convergence expected at the sequence level. Accordingly, convergence does not begin to undermine the assumptions of this analysis until much larger values of λt (around 0.15) than those for restriction sites are found. The domain of applicability of this nonparametric analysis is therefore somewhat broader for sequence data than for restriction site data.

ACKNOWLEDGMENTS

I thank Dr. Joseph Felsenstein for his valuable suggestions in the preparation of this chapter. This work was supported by Grant Number GM27021 from the National Institutes of Health.

Chapter 8

THE NUMBER OF POLYMORPHIC DNA CLONES REQUIRED TO MAP THE HUMAN GENOME

D. Timothy Bishop

Department of Human Genetics, University of Utah, Salt Lake City, Utah

Christopher Cannings

Department of Probability and Statistics, University of Sheffield, Sheffield, England

Mark Skolnick

Department of Human Genetics, University of Utah, Salt Lake City, Utah

John A. Williamson

Department of Mathematics, University of Colorado, Boulder, Colorado

I. INTRODUCTION

A major endeavor in human genetics is the determination of the mode of inheritance of genetic traits. In particular, the identification of a locus of major effect improves the ability of genetic counsellors to assign risks to relatives of affected individuals.

The presence of a major locus also suggests that a substantial contribution to the etiology of a given disease can be traced to a single biochemical defect. Evidence for major gene involvement in the determination of a genetic defect is greatly strengthened if it is possible to demonstrate linkage between the major locus and a marker locus of known inheritance and chromosomal location. Chromosomal localization of a major gene can also lead to etiological hypotheses, and a discussion of the use of genetic markers for linkage and association studies is given in the chapter by Clegg and Asmussen in this volume. The benefit to evolutionary studies of genetic markers defined by DNA sequence variation is discussed in the chapter by Felsenstein.

The ability to localize major genes on human chromosomes is primarily dependent upon the availability of marker loci. Two factors determine the utility of marker loci for linkage studies. Firstly, marker loci must be spread out over the genome as no major locus can be assigned to a region that does not contain a marker. Secondly, the marker loci should have a large number of alleles, with no allele so common that homozygotes occur frequently. The consequent high number of segregating families increases the power of detection of linkage, and this criterion explains the great value of the HLA loci for genetic linkage studies. The equivalent of multiple allelism can also be achieved by considering several linked marker loci as a single marker of a given region. In this chapter we will discuss the first factor, the location of marker loci, in great detail. In particular, we will consider the number of randomly identified marker loci required to ensure that, with high probability, any disease locus is within a given recombination distance of at least one marker locus. Refinements relating to the degree of polymorphism are considered by Bishop and Skolnick (1983).

Marker loci reflect DNA sequence variation between individuals and such variation has usually been assayed as isoenzyme or antigenic polymorphism. The discovery of a DNA polymorphism associated with the β-globin gene cluster (Kan and Dozy, 1978a), and the study of similar systems in other eucaryotes, such as yeast (Goodman *et al.*, 1977; Petes and Botstein, 1977; Cameron *et al.*, 1979), lead to the proposal that arbitrary cloned DNA segments could be screened to reveal restriction fragment length polymorphisms, or RFLP's (Botstein *et al.*, 1980). These RFLP's could be used as markers, and Botstein *et al.* (1980) proposed that they be sought from clones chosen at random from a human DNA library. Such a library could be either total DNA, or DNA complementary to messenger RNA, or DNA from a human-rodent hybrid (Gusella *et al.*, 1981) or DNA enriched for a specific chromosome by sorting (Davies *et al.*, 1981) or sedimentation. The search would continue until a sufficient number of RFLP's had been found. Botstein *et al.* (1980) argued that, based on the difficulty of detecting loose linkage (Thompson *et al.*, 1978), the gene under investigation should not be further than 20 centiMorgans (cM) from the nearest marker locus, and that, therefore, a set of markers 40 cM apart and evenly spread over the genome are needed. (Two loci that have a genetic map distance of one Morgan between them are expected to have one crossover between them at meiosis. A map distance of one centiMorgan is called a map unit, and this is to be distinguished from the physical distance between the loci.) A similar conclusion was reached in Bishop and Skolnick (1980), based on the expected number of electrophoretic gel slots required to map a major locus when a set of evenly spaced chromosome markers is available.

Using Renwick's (1971) estimate of 33 Morgans for the map length of the human genome, Botstein *et al.* (1980) concluded that a set of 83 markers, 40 cM apart, should be selected for an efficient primary linkage set. In fact, since the markers are selected at random from a given DNA library, the number of markers required to achieve the desired goals is considerably more than 83.

The figure of 40 cM contains an inherent paradox. It is calculated on the assumption that the maximum distance between any locus and the closest marker should be, at most, 20 cM. The efficient set of 83 markers is then taken from the much larger set of randomly chosen markers but in order to be able to select such a subset, it must be possible to identify the linkage groups amongst the markers themselves, with the consequent requirement that adjacent *markers* are, at most, 20 cM apart.

The number of markers required to cover the genome may depend on the strategy employed to find the RFLP's. This chapter will consider in detail two strategies. The first involves the use of a genomic library while the second uses single chromosome libraries. From this second type of library, markers can be selected to avoid choosing new probes for a chromosome which already carries sufficient markers.

In this chapter, we will use both analytical results and computer simulations to calculate the number of markers required, on the average, to map a given proportion of the human genome. Aspects of the problem have also been considered by Elston and Lange (1975) and by Lange and Boehnke (1982). Our mathematical discussion will be followed by tabulated results for parameter values relevant to the human genome. Our interpretation of these results will emphasize how factors we have not considered may affect the analysis. For brevity in this discussion we refer to the autosomal genome as simply the genome. We also present our estimate of the effective number of markers required to map the human genome.

II. MATHEMATICAL ANALYSIS

In this section we present the basic mathematical analysis. The development appears in the mathematical literature under the general heading of solutions to various forms of the "covering problem." The basic covering problem considers placing random arcs of equal length on a circle of unit circumference. Solutions to the problem take many forms. In the first place, the density of the proportion covered, the expected proportion covered, and other moments have been derived for the case where the number of random arcs placed is a fixed integer, n. In the second place, the number, N, of randomly placed arcs needed to completely cover the whole, or some prescribed portion, of the circumference is considered. Random arcs are placed until the desired coverage is achieved, and N is a random variable whose distribution, mean and variance for the complete covering have been obtained by many authors (*e.g.*, Flatto and Konheim, 1962). Such results may be adapted to the case of a line segment, and it is this situation that is of interest for the mapping of the human genome.

A. Model and Notation

In the context of mapping the human genome, the line segment represents the *linkage* map of a single chromosome from a diploid population. This chromosome will be taken to be t cM in map length. A polymorphic marker locus is identified experimentally, and randomly in the sense that such loci are regarded as being distributed uniformly on the *linkage map*. Finding such a locus is thus equivalent to choosing a random point on a line segment of length t. Any major, or disease, locus close to this polymorphic marker can be identified through linkage studies (Morton, 1955). We suppose that if the distance between the marker locus and the locus of interest is less than x/2 cM, then linkage will be shown. The interval of length x cM centered on the marker contains all the disease loci which can be shown to be linked to the marker, and the chromosome is considered covered if every disease locus is closer than x/2 cM to at least one marker locus. Generally, to be able to show linkage with a reasonable number of human pedigrees, x should be less than 40 cM.

B. Probability of Complete Cover with n Markers

Suppose that n marker loci have been identified on a single chromosome of map length t cM. The probability that the whole of the chromosome is covered by the n intervals of length x cM centered at the marker loci is c(n,t,x), where

$$c(n,t,x) = \sum_{v=0}^{n-1} (-1)^v \binom{n-1}{v} \{(1 - \frac{vx}{t})_+^n - 2[1 - (v + \tfrac{1}{2})\frac{x}{t}]_+^n + [1 - (v+1)\frac{x}{t}]_+^n\} \quad (1)$$

and

$$a_+ = \max(a,0)$$

Stevens (1939) obtained the corresponding result for random arcs placed on the circumference of a circle. Equation (1) follows from the work of Stevens by conditioning on the midpoint of the extreme intervals on the line, and then treating the other (n-2) intervals as being on the circle created by joining the midpoints of the extreme intervals.

C. Moments of Proportion of Genome Covered with n Markers

The mean and variance of the proportion C_n covered by n markers randomly placed on r chromosomes can be derived from standard manipulations of the density function but are more easily obtained from Robbins' theorem (Robbins, 1944, 1945). This theorem allows us to find the expected coverage $E(C_n)$ in terms of the probability $p_{1,n}$ that a randomly placed point is not covered by n randomly placed intervals:

$$E(C_n) = 1 - p_{1,n} \quad (2)$$

Further, if $p_{2,n}$ is the probability that neither of two randomly placed loci is covered by n randomly placed intervals, then

$$E[(1-C_n)^2] = p_{2,n}$$

and

$$\text{var}\ (C_n) = p_{2,n} - p_{1,n}^2$$

We consider r chromosomes of lengths $t_1, t_2, \ldots, t_r$ cM ($\Sigma t_i = t$), with n intervals of length x placed on the genome in such a way that their centers are placed randomly and independently over the total map length. The derivation of $p_{1,n}$ requires the consideration of two cases. In the first place, a random locus falls in the interior of a chromosome, at least x/2 cM from each end, with probability (1 - rx/t). The probability that this locus is not covered by any of the n intervals is

$$(1 - x/t)^n$$

A correction for end effects constitutes the second case. Now the locus falls at a distance $y < x/2$ cM from an end of the chromosome map, and the probability that such a point is not covered is

$$[1 - (y + x/2)/t]^n$$

Multiplying this by xr/t, the density function of y, integrating over $0 < y < x/2$, and adding the result to the above result, gives

$$p_{1,n} = \frac{2r}{n+1} \left[\left(1 - \frac{x}{2t}\right)^{n+1} - \left(1 - \frac{x}{t}\right)^{n+1} \right] + \left(1 - \frac{rx}{t}\right)\left(1 - \frac{x}{t}\right)^n \quad (3)$$

This equation requires that $t_i > x$ for all i, a condition which is satisfied for the human genome since the shortest autosome has map length about 55 cM and the values of x that are of interest are less than 40 cM.

Elston and Lange (1975) obtained equation (3) for the case of r = 1, and extended the result to r > 1 by conditioning on the number of markers falling on each chromosome and considering all possible arrangements. They did not simplify the resulting summation by requiring $t_i > x$ for all i.

The probability $p_{2,n}$ can be calculated in a similar fashion. If the assumption is made that each chromosome is at least 2x cM in map length (as is generally the case for situations of interest in this chapter), then

$$p_{2,n} = \frac{2r}{(n+1)(n+2)} [\phi(2)^{n+2} - 2\phi(3)^{n+2} + \phi(4)^{n+2}] + \frac{4r}{n+2} [\phi(3)^{r+2}$$

$$- \phi(4)^{n+2}] - \frac{4(r-1)}{n+1} [\phi(3)^{n+1} - \phi(4)^{n+1}] + \frac{4r}{(n+1)(n+2)} [\phi(1)^{n+2}$$

$$- \phi(2)^{n+2} - \phi(3)^{n+2} + \phi(4)^{n+2}] + \frac{2r}{n+2} [\phi(2)^{n+2} - \phi(4)^{n+2}]$$

$$- \frac{2(r-1)}{n+1} [\phi(2)^{n+1} - \phi(4)^{n+1}] + \frac{4r(r-1)}{(n+1)(n+2)} [\phi(2)^{n+2} - 2\phi(3)^{n+2}$$

$$+ \phi(4)^{n+2}] + \frac{4(r-1)}{n+1} (1 - \frac{rx}{t}) [\phi(3)^{n+1} - \phi(4)^{n+1}]$$

$$+ (1 - \frac{2x}{t})^n [(1 - (r+1) \frac{x}{t})^2 + (r-1) \frac{x^2}{t^2}] \tag{4}$$

where

$$\phi(k) = (1 - \frac{kx}{2t}) .$$

Note that, under the assumption of each chromosome being at least 2x cM in map length, $p_{1,n}$, $p_{2,n}$ and $\text{var}(C_n)$ are all independent of the t_i's, but dependent on t and r. For the special case of a single chromosome (r = 1) of length t,

$$p_{1,n} = \frac{2}{n+1} [\phi(1)^{n+1} - \phi(2)^{n+1}] + \phi(2)^{n+1} \tag{5}$$

$$p_{2,n} = \frac{1}{(n+1)(n+2)} [4\phi(1)^{n+2} + 2n\phi(2)^{n+2} + 4(n-1)\phi(3)^{n+2} + (n-1)(n-2)\phi(4)^{n+2}] \tag{6}$$

The reason for the dependence of $p_{1,n}$ and $p_{2,n}$ on t is clear, while the dependence on r arises from the fact that coverage is lost whenever a marker locus falls near the end of a chromosome. The larger the number of chromosomes, the smaller will be the expected coverage. For $r = 0$, i.e., no ends, equations (2) and (3) correspond to the results of Solomon (1978) for the circle.

When n markers have been located, we know that the expected coverage of the genome is the complement of $p_{1,n}$, given by (2). The expected coverage of the *ith* chromosome, $E[C_n(t_i)]$ is then, presumably, a function of its length. The value is found by considering the number, m, of markers that fall on that chromosome and then weighting the expected proportion covered by m markers [equation (5) with n replaced by m] by that probability. The result is

$$E[C_n(t_i)] = (1 - \frac{x}{t_i})(1 - \frac{x}{t})^n + \frac{2t}{(n+1)t_i} [(1 - \frac{x}{2t})^{n+1} - (1 - \frac{x}{t})^{n+1}] \tag{7}$$

D. Moments of Numbers of Markers Needed for Partial Covering

We now consider the number of intervals, N(b), required to cover a proportion b of a line segment of length t. For $b = 1$, $\mathrm{Prob}[N(1) \le n]$ is the probability that the whole segment is covered with n, or fewer, intervals and this probability is given by equation (1). The expected value of N(1) is

$$E[N(1)] = \sum_{n=0}^{\infty} \{1 - \mathrm{Prob}[N(1) \le n]\}$$

and the variance is

$$\mathrm{var}\ [N(1)] = 2 \sum_{n=0}^{\infty} n\{1 - \mathrm{Prob}[N(1) \le n]\} + E[N(1)] - E^2[N(1)]$$

Expressions for these moments have been given by Cooke (1974). By interchanging the order of summation and using various forms of the identity

$$\sum_{n=k}^{\infty} \binom{n}{k} p^{k+1}(1-p)^{n-k} = 1 \qquad \text{for} \qquad 0 < p < 1$$

it is possible to obtain alternative forms for the moments which are more stable numerically. These alternatives are

$$E[N(1)] = 1 + \sum_{v=1}^{\infty} (-1)^{v+1} \left(\frac{t}{vx} - 1\right)_{+}^{v+1} - 2 \sum_{v=0}^{\infty} (-1)^{v+1} \left[\frac{t}{(v+\frac{1}{2})x} - 1\right]_{+}^{v+1} + \sum_{v=0}^{\infty} (-1)^{v+1} \left[\frac{t}{(v+1)x} - 1\right]_{+}^{v+1} \quad (8)$$

$$\mathrm{var}[N(1)] = E[N(1)] - 2\Big[\sum_{v=1}^{\infty} (-1)^{v} \frac{t(v+1)}{vx} \left(\frac{t}{vx} - 1\right)_{+}^{v+1} - 2 \sum_{v=0}^{\infty} (-1)^{v} \frac{t(v+1)}{(v+\frac{1}{2})x} \left[\frac{t}{(v+\frac{1}{2})x} - 1\right]_{+}^{v+1} + \sum_{v=0}^{\infty} (-1)^{v} \frac{t}{x} \left[\frac{t}{(v+1)x} - 1\right]_{x}^{v+1}\Big\} - E^2[N(1)] \quad (9)$$

For the case of r chromosomes of lengths $t_1, t_2, \ldots, t_r$ $(\Sigma t_i = t)$

$$E[N(1)] = \sum_{n=0}^{\infty} \{1 - \mathrm{Prob}[N(1) \leq n]\} \quad (10)$$

where

$$P[N(1) \leq n] = \sum_{k_1,k_2\ldots k_r} \frac{n!}{k_1!k_2!\ldots k_r!} \prod_{i=1}^{r} \left(\frac{t_i}{t}\right)^{k_i} c(k_i,t_i,x)$$

$$\Sigma k_i = n, \quad k_i \geq 0$$

and $c(k,t,x)$ was defined in equation (1). Approximate values of this expectation, using $r = 22$ for the human genome, have been obtained by simulation by Lange and Boehnke (1982).

E. Size of Uncovered Regions

It is clear from simulation and numerical results based on equations (8) and (9) that a complete covering of the chromosomes takes many more random intervals than does, say, a covering of 90% of the chromosomes. A few small uncovered regions require a large number of intervals in order that the covering may be completed.

It is of interest then to consider the distribution of gap lengths in a partial covering, with proper attention being paid to the practical importance of gaps. When x = 20 cM, a gap of 2 cM would give a distance between marker loci of 22 cM. While there is a gap in the mathematical sense, in practice it may be of little consequence since there is not a discontinuity in mapping ability at map distances of x/2 from each marker.

Our first approach is numerically and analytically simple. We say that an h-partial covering is achieved if none of the gaps between the covered regions exceeds h cM. If end effects are neglected, this is equivalent to saying that replacing intervals of length x by intervals of length x+h totally covers the line segments. The expected number of intervals which must be placed to achieve an h-partial covering will be less than or equal to the expected number of intervals of length x+h needed for a complete cover. The inequality is a consequence of end effects.

The second approach is to consider the size of gaps when a given number of marker loci have been placed. Our results will be based on simulations.

F. Simulations

Our simulation program allows two options. In one, covering continues until a fixed proportion of the library is covered, and, in the other, a fixed number of intervals are placed randomly. The library can be taken to be the human autosomal chromosome complement of 22 chromosomes, or just a single chromosome. In either case, the lengths of the individual chromosomes are taken to be proportional to the values shown in Table 2. Tabulated results are based upon 1000 simulations.

G. Approximation

Several of the formulae relating to the covering and partial covering problems are not computationally feasible, and care must be taken in evaluating others in some circumstances. As an approximation, the model without end effects may be used with a fair degree of confidence. Approximate treatments can also be based on the classical occupancy problem (e.g., Feller, 1968).

III. RESULTS

In this section we apply the methodology of the previous section. We assume that the human autosomal genome is of total map length t = 33 Morgans. This figure was obtained by Renwick (1971) as a sex-averaged estimate. The figures for males and females were 27.5 and 38.5, respectively. As these values are in some doubt (Hulten *et al.*, 1981), we consider values other than 33.

The expected proportions of the genome covered when n RFLP's have been identified are shown in Table 1. The values were computed from equations (2), (3) and (4), except for the standard deviations in the cases of x = 40 cM and t = 25 or t = 33 Morgans. In those situations, some individual chromosomes are less than 2x cM in map length and equations (3) and (4) are only approximate. Simulations are then used to provide standard deviations.

Table 1. Expected proportion (with standard deviation) of genome mappable with n randomly chosen markers. [Based on equations (2), (3) and (4) with r=22 chromosomes.]

		n				
t	x	50	100	200	400	800
25	20	0.32(0.02)	0.53(0.02)	0.78(0.02)	0.95(0.01)	1.00(0.00)
	40	0.52(0.02)	0.76(0.03)	0.94(0.02)	0.99(0.00)	1.00(0.00)
33	20	0.25(0.01)	0.44(0.02)	0.69(0.02)	0.90(0.01)	0.99(0.00)
	40	0.43(0.02)	0.67(0.03)	0.89(0.02)	0.99(0.01)	1.00(0.00)
40	20	0.22(0.01)	0.39(0.01)	0.62(0.02)	0.85(0.01)	0.98(0.01)
	40	0.38(0.02)	0.61(0.02)	0.84(0.02)	0.97(0.01)	1.00(0.00)

The expected coverage is calculated for x = 20 cM and for x = 40 cM. If x = 20, the markers are close enough so that linkage between adjacent markers can be shown. In this case, the figures in Table 1 indicate the proportion of the genome which is covered sufficiently well that an efficient subset of markers can be chosen for linkage studies. When x = 40, it may not be possible to show linkage between the markers, but the figures in the table indicate the proportion of the genome that falls within a distance of x/2 = 20 cM of the closest marker. Any disease locus within this cover can be shown to be linked to at least one marker locus. The table also includes three values of t to represent hypothesized values of the human autosomal map length. We see that for n = 50 the expected proportion covered is an appreciable percentage of the maximum possible covered proportion of 50x/t. When t = 33 Morgans, for example, the expected proportion covered if x = 20 cM is 0.25, while the maximum possible coverage is 0.31. This indicates how an initial random search of the whole genome is efficient in terms of coverage achieved with few marker loci. As the genome becomes saturated, however, a random search becomes highly inefficient in that it duplicates coverage, as can be seen from examining the proportions covered for n values of 200 or greater.

We next consider the number of markers, N(1), required for a total covering, with every locus within a map distance of x/2 cM from at least one polymorphic locus. This complete covering may be achieved either by covering each chromosome before moving on to the next, or by randomly searching the genome until there are no uncovered regions. We first consider the individual chromosome library.

The work of Davies *et al.* (1980), Gusella *et al.* (1980) and others allows chromosome-specific libraries to be made. With such libraries it is possible to find markers on each chromosome, one chromosome at a time. In Table 2 we show the expected number of markers required to cover each chromosome. The calculations, based on equations (8) and (9), require the map length of each chromosome and we assume that this is proportional to the DNA content of that chromosome. Bosman *et al.* (1977) obtained the DNA content of each chromosome, averaged over five normal individuals, using scanning densitometry. We use their values, and see in Table 2 that, with t = 33 Morgans and x = 20 cM, the expected number of markers required to cover chromosome 1 is 74.9, while equation (2) provides that the expected proportion of chromosome 1 covered by 20 markers is 0.76 and by 40 markers is 0.94. Summing the expected values in Table 2 gives the expected number of markers required to cover the genome using single chromosome libraries.

Table 2. Expected number (with standard deviation) of number of markers required to cover each chromosome separately. [Based on equations (8) and (9).]

Chromosome	Map length (Morgans)	x=20 (cM)	x=40 (cM)
1	2.88	74.9 (24.4)	32.5 (12.5)
2	2.77	71.4 (23.5)	31.0 (12.1)
3	2.30	57.2 (19.7)	24.7 (10.1)
4	2.16	53.0 (18.5)	22.9 (9.2)
5	2.11	51.5 (18.1)	22.2 (9.2)
6	1.94	46.6 (16.7)	20.1 (8.5)
7	1.80	42.5 (15.5)	18.3 (7.9)
8	1.66	38.6 (14.4)	16.6 (7.3)
9	1.61	37.2 (14.0)	16.0 (7.1)
10	1.55	35.5 (13.4)	15.2 (6.8)
11	1.55	35.5 (13.4)	15.2 (6.8)
12	1.52	34.7 (13.2)	14.9 (6.7)
13	1.22	26.5 (10.7)	11.4 (5.3)
14	1.14	24.4 (10.0)	10.5 (5.0)
15	1.11	23.7 (9.7)	10.1 (4.8)
16	1.05	22.1 (9.2)	9.5 (4.6)
17	1.00	20.8 (8.8)	8.9 (4.3)
18	0.92	18.8 (8.1)	8.0 (4.0)
19	0.75	14.6 (6.6)	6.3 (3.2)
20	0.78	15.4 (6.8)	6.6 (3.3)
21	0.55	10.0 (4.8)	4.3 (2.2)
22	0.61	11.4 (5.3)	4.9 (2.5)
Totals	33.00	766.3	330.1

The results of such summations are contained in Table 3 under the entry for t = 33 Morgans and b = 1.0. A similar procedure was adopted for other "chromosomal" entries in Table 3. For comparison, Table 3 also contains the expected number of markers required to map the genome with a genomic library. Such values were obtained by simulation since the analytical approach [equation (10)] is not numerically tractable. Results for t = 33 and b = 1.0 similar to those in Table 3 are given by Lange and Boehnke (1982).

Table 3. Expected number (with standard deviation) of markers required to cover genome to at least a proportion b, with genomic or chromosomal libraries.

			b			
t	x	Library	0.5	0.7	0.9	1.0
25	20	genomic	91 (5)	158 (9)	305 (19)	1130 (284)
		chromosomal	99 (5)	162 (10)	302 (22)	548 (50)
	40	genomic	48 (4)	84 (7)	161 (15)	512 (115)
		chromosomal	59 (4)	88 (7)	151 (14)	235 (25)
33	20	genomic	118 (6)	206 (9)	397 (22)	1561 (386)
		chromosomal	126 (6)	213 (11)	396 (26)	766 (66)
	40	genomic	62 (5)	107 (7)	206 (16)	744 (182)
		chromosomal	71 (4)	111 (8)	200 (17)	330 (33)
40	20	genomic	142 (6)	247 (11)	470 (23)	1725 (270)
		chromosomal	151 (7)	252 (13)	465 (28)	967 (79)
	40	genomic	73 (4)	128 (9)	243 (19)	886 (217)
		chromosomal	82 (5)	133 (9)	237 (19)	417 (41)

As expected, a comparison of genomic and chromosomal entries in Table 3 shows that single chromosome libraries require substantially fewer markers to cover the genome completely. With x = 40 cM and t = 33 Morgans, for example, approximately 1561 markers are needed with a genomic library approach, while only 766 are needed with a single chromosome library approach. Disregarding the extra effort in constructing single chromosomal libraries, it appears that this method is about twice as efficient. Presumably it would be possible to increase efficiency further by using other fractions of the genome, or by switching from a genomic to a chromosome-specific library at an appropriate stage.

We now examine the number of markers required for the map to be effectively complete. The previous discussion has focused on complete coverage of the genome, but it is apparent from Table 1 that a large proportion of the genome can be mapped with a small fraction of the number of markers required for complete coverage. In Table 3 we show the number of markers required to map a proportion, b, of the genome using either genomic or chromosomal libraries. For $b < 0.9$, the expected numbers of markers under the two strategies are almost identical. Other criteria may cause some differences, however, and two factors in particular warrant attention.

Firstly, when each chromosome is covered to at least a proportion b, the final cover of the genome will be more than b. With $b = 0.5$, $x = 20$ cM and $t = 33$ Morgans, for example, approximately 55% of the genome will be covered. For a genomic library, the expected number of markers required to cover 55% of the genome, $E[N(0.55)]$, is approximately 136 and $E[N(0.5)]$ is 118, while, using chromosomal libraries, a 50% cover is achieved with 126 markers. The additional cover gained from seeking a 50% cover of the genome does not, by itself, appear to justify the use of the chromosomal library.

Secondly, covering each chromosome to a given proportion, b, may map the genome more effectively than covering the whole genome to the same proportion, even after accounting for the increased cover from the individual chromosome libraries. For example, if the uncovered regions were more numerous but smaller from chromosomal libraries, this strategy might be considered more efficient. Each chromosome will be 90% covered with 385 markers and chromosomal libraries, while this same number randomly drawn from the whole genome will give an 89% overall coverage. In Table 4 we show the distribution of uncovered regions for both types of library when $x = 20$ cM and $t = 33$ Morgans. The numbers of uncovered regions are approximately equal (44 for genomic and 42 for chromosomal libraries). The uncovered regions are slightly smaller (7 cM) on average for chromosomal than (8 cM) for genomic libraries, but, again, the difference does not appear to be significant.

Table 4. Average number of gaps for genomic and chromosomal approaches. (Based on simulations with t=33, x=20 and n=385.)

Size of uncovered region (centiMorgans)	Library Genomic	Chromosomal
0 - 4	20.2	20.6
5 - 9	10.6	10.8
10 - 14	6.1	5.5
15 - 19	3.2	2.8
over 19	3.6	2.7

Earlier we suggested that ignoring end effects may lead to a useful approximate treatment. Indeed, if we set $r = 0$, so that the genome is assumed to be circular, equation (3) provides values for Table 1 that are within 0.02 of the exact values. Figures produced for $r = 0$ always exceed those for $r > 0$ since coverage is always lost when intervals overlap chromosome ends. Lange and Boehnke (1982) have considered the expected number of markers required to cover a circular genome using the formula of Flatto and Konheim (1962). They show that this quantity seriously underestimates the number required for separated linear chromosomes. This latter value was obtained by simulation. As analytical results are also limited in applicability for a circular genome, this approximation is of little utility for future progress.

IV. DISCUSSION

We suggest that, based on the previous results and the following rationale, approximately 400 markers will suffice to cover the human genome and provide a subset of RFLP's for linkage studies. Table 3 indicates that, for markers to be within 20 cM of each other with probability 0.9 so that a primary set of markers can be selected for linkage studies, slightly fewer than 400 markers are required. At this marker density, Table 4 indicates that there will be between three and four gaps of sufficient size (greater than 20 cM) that a disease locus in the gaps will be undetectable. On the average, therefore, after covering 90% of the genome, three or four chromosome specific libraries will be required to complete the map. Such a combined strategy is appropriate when the cost of making chromosome-specific libraries is considered (Cannings, in preparation). With 400 markers, equation (7) shows that chromosome 1 is 91% covered, so that about 27 cM of map length is uncovered. For chromosome 21, the uncovered regions amount to about 7 cM. Simulations show that the largest chromosome has a probability of approximately 0.2 of containing at least one gap of 20 cM. For chromosome 21, the corresponding probability is less than 0.05.

Specific strategies can be utilized to cover those regions of the genome that have RFLP's more than 20 cM from each other. The region covered by an RFLP is dependent upon the number of alleles available at the locus and on the frequency of those alleles. The region can be extended by increasing the number of identifiable alleles, and hence decreasing homozygosity. The first random marker locus identified with recombinant DNA technology has been shown to have a higher degree of polymorphism than all known markers except HLA (Wyman and White, 1980), when polymorphism is defined as the probability that an offspring of a parent carrying a rare allele at a disease locus will allow deduction of the parental genotype (Botstein *et al.*, 1980). Large gaps on a chromosome will be covered following chromosome-specific searches. A chromosome will be considered covered when a single linkage group of sufficient size exists. For chromosomes where the markers form several linkage groups, the RFLP's at the ends of each group can be made more informative to see if linkage between the groups can be established. If this fails more probes must be sought.

Although we suggest that 400 markers are required, Table 1 shows that there is a probability of nearly 0.9 that any disease locus will be mappable with 200 markers. In this case the map distances between all markers will not be known.

The model we have used contains a number of assumptions that may affect the actual number of markers required. More than 400

testify that this implies latent homosexual urges?

3. Call Fred's Old Friend.

4. We need advance notice of when Fred will be notified of any legal failure by his lawyer so we can leave town.

1. I documented case of a ~~phys~~ shrink giving him a clean bill of health 5/23/83 one month before suicide attempt & involuntary commitment. 6/2/83

2. Documentation of Fred's excessive Fear of homosexuality.

will be needed, for example, if the total map length is more than 33 Morgans.

The second crucial assumption is that map length is proportional to physical distance. If map distance is not uniform over the genome then the mathematical results are still approximately true provided that increased recombination is associated with higher polymorphism. This increased polymorphism may exist naturally or be created experimentally, for example, through the use of several restriction enzymes with a single probe. Regions of increased recombination (*e.g.*, telomeric regions) that do not have increased polymorphism will be physically underrepresented in the library and will be difficult to map. Our estimates would be too low in these circumstances.

Finally, clones are represented in the libraries in proportion to their physical lengths and the propensity of those segments of DNA to be cloned. Therefore, the amount of effort required will also be increased if DNA from particular regions is difficult to clone (*e.g.*, highly repetitive regions). Such regions may require the development of specific strategies.

In practice, the aims of diverse investigators may not be served by the strategies designed to map the genome efficiently. If the strategy of this chapter was followed in a collaborative way by investigators around the world, genomic libraries would be considered until enough chromosomes were covered to indicate a switching to single chromosomal libraries (Cannings, in preparation). At such a time, small chromosomes would tend to be covered and large ones partly covered. If chromosome-specific searches were to be initiated before this appropriate switching time, attention should therefore be focused on the larger chromosomes. In fact, other considerations have led to smaller chromosomes. In cially number 21, receiving greater attention in several laboratories. Because RFLP's are also created from cloned genes, the total genomic search will continue after the strategy suggested for our purpose would indicate a switch to single chromosome libraries.

These considerations lead us to believe that certain areas of the genome will become highly saturated, increasing the number of markers required until the total genome is covered. By June of 1981 there were 25 RFLP's known (Skolnick and Francke, 1982) and this figure had at least doubled a year later. At such a rate, the genome will be effectively covered by 1985.

ACKNOWLEDGMENTS

We wish to thank Kenneth Lange and Michael Boehnke for making available to us a prepublication copy of their paper and for indicating many of the mathematical references to the covering problem. We also thank Ray White for many discussions related to DNA polymorphisms, and Joan Brewer and Chris Hall for typing a draft of this chapter. This research was supported by grants CA 28854 and GM 27192 from the National Institutes of Health.

Chapter 9

USE OF RESTRICTION FRAGMENT POLYMORPHISMS AS GENETIC MARKERS

Michael T. Clegg

Departments of Molecular and Population Genetics, and Botany
University of Georgia
Athens, Georgia

Marjorie A. Asmussen

Departments of Molecular and Population Genetics, and Mathematics
University of Georgia
Athens, Georgia

I. INTRODUCTION

Marker genes have long been employed to dissect the genetic architecture of major phenotypic traits (*e.g.*, Mather and Jinks, 1971). Indeed, it is easy to imagine a variety of ways in which such genes could be employed in areas ranging from plant and animal breeding through to human genetics. These applications include constructing detailed linkage maps (see chapter by Bishop *et al.*, in this volume), genetic analysis of complex phenotypic traits, and predicting the transmission of genes associated with the marker locus in specified matings. This latter application is especially important in medical genetics where prediction of genetic pathologies is desired. The major factor limiting the use of marker genes has been the relatively small number of polymorphic loci, in most species, which could serve as genetic markers. This obstacle may soon be overcome, however, by the analysis of genetic differences at the level of DNA sequences. Such studies have estimated that the probability that a given nucleotide site will be polymorphic ranges from 0.001 to 0.01 (Botstein *et al.*, 1980). Variation at this level therefore promises to provide a virtually unlimited source of new genetic markers.

One relatively straightforward means of detecting these nucleotide sequence polymorphisms involves the use of type II restriction endonucleases together with cloned DNA sequences and Southern (1975) transfers. Briefly, type II restriction endonucleases cleave double stranded DNA molecules wherever a specific permutation of nucleotides occurs (the recognition site). For example, the restriction endonuclease *Eco* RI has the recognition sequence GAATTC. *Eco* RI digestion of a DNA molecule yields a population of fragments whose termini are defined by recognition sites for *Eco* RI. The fragments so obtained can be size fractionated by gel electrophoresis (see chapter by Schaffer in this volume). A specific cloned DNA sequence can then be used as a hybridization probe to permit the direct visualization of only those fragments carrying sequences homologous to the probe. Nucleotide substitutions may alter the distribution of fragment sizes by either destroying a previously existing recognition site, or by creating a new one. Fragment sizes may also be modified by insertions, deletions and rearrangements of DNA sequences. Thus, differences in the distribution of fragment patterns among individuals indicate underlying DNA sequence polymorphisms. Such polymorphisms are referred to as restriction fragment length polymorphisms (RFLPs).

Beyond the potentially large number of RFLPs in populations of higher organisms, two other advantages of this class of markers are (i) codominance in expression, and (ii) direct physical mapping of genes and closely associated restriction site polymorphisms.

The possibility of direct physical mapping follows from the fact that the DNA probe may be found on fragments of two or more lengths in a population. The distance in physical units (such as kilobase pairs, kbp) from the variable site to the gene can be easily measured. More involved techniques such as "DNA walking" can also be employed for physical mapping when the distance between the polymorphic restriction site and the gene of interest is too great to be spanned by a single fragment. The cases of very tight linkage, to which the DNA based techniques apply, however, are precisely those where the estimation of recombination fractions from crosses is most tedious. The beauty of DNA procedures is that they circumvent the need to make genetic crosses and to study genetic transmission through families.

The purpose of this chapter is to evaluate the use of RFLPs as genetic markers in predicting the transmission of pairs of genes in certain kinds of matings. We first consider the transmission of associated genes at linked loci. Next we investigate how the values of the population variables (gene frequencies and linkage disequilibrium) influence the class of matings in which the marker gene provides information on the transmission of the associated gene. We then consider the evolutionary dynamics of the population variables to determine their most likely states under different model assumptions. Finally, we discuss various conditional measures of association between genes at a target locus and those at a linked marker locus.

II. THE PROBLEM

We assume two diallelic autosomal loci A and M where M is the marker locus and A is the associated, or target, locus. The recombination fraction between loci A and M is denoted by r. Locus A may specify any attribute; however, for purposes of discussion we will assume that an allele of A, say A_2, determines a genetic defect when in homozygous form. The objective is then to use the transmission of alleles at locus *M* as indicators of the transmission of particular alleles at locus *A*. As a concrete illustration we consider the case of sickle cell anemia, which results from a particular amino acid substitution at the sixth position from the N-terminal end of the β-globin polypeptide. Kan and Dozy (1978a) reported a *Hpa* I restriction site polymorphism adjacent to the β-globin locus in human populations and they subsequently used this polymorphism for prenatal diagnosis of the genotype at the β-globin locus (Kan and Dozy, 1978b). Figure 1 gives a diagrammatic representation of this RFLP associated with the β-globin locus. In terms of our established notation, the sickling allele is denoted A_2 and we will denote the *Hpa* I site which produces the 7.0 (or 7.6) kbp fragment by M_2. The absence of this site results in a 13.0 kbp fragment denoted by M_1. The physical distance between M_2 and the 3' end of the β-globin locus is approximately 5,000 base pairs and r is estimated to be 0.00005 (Kurnit and Hoehn, 1979).

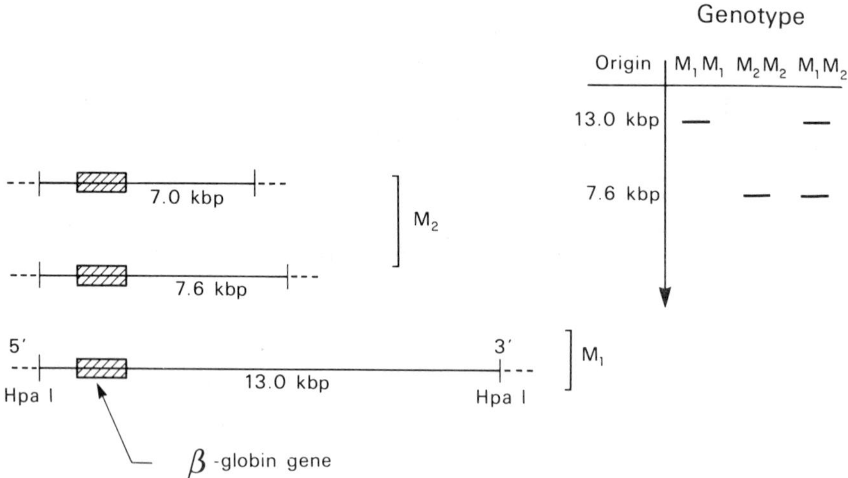

Figure 1. Diagrammatic representation of the polymorphic restriction fragments associated with the β-globin locus. The β^S allele is known to be strongly associated with the 13.0 kbp fragment (allele M_1) in human populations of West African origin. The 7.0 and 7.6 kbp fragments will be treated as a single allelic class (allele M_2) for purposes of discussion. The fragment mobilities associated with the three marker locus genotypes are also depicted.

III. TWO LOCUS MATINGS

The likelihood of confirming or excluding the A_2A_2 genotype on the basis of the marker genes transmitted hinges strongly on the degree of linkage between the two loci. Consider, for example, the mating of the double heterozygotes A_1M_2 / A_2M_1 by A_1M_2 / A_2M_1. Of the progeny resulting from this mating, 1/4 will have the genotype M_1M_1 and $(1-r)^2$ of these will also be A_2A_2. The remaining two genotypic classes will be A_1A_1 or A_1A_2 with probability $1 - (1 - r)^2$. The odds that the occurrence of M_1M_1 correctly predicts the genotype A_2A_2 are thus $(1 - r)^2 / [1 - (1 - r)^2]$, or, for small r, approximately 1/2r (*i.e.*, 10,000:1 in the sickle cell case). If, on the other hand, r is 0.1, the odds of M_1M_1 correctly predicting A_2A_2 are only about 4.2:1. It is also possible that the marker genotypes M_1M_2 or M_2M_2 could be associated with the deleterious A_2A_2 genotype. The conditional probability of A_2A_2 given M_1M_2 or M_2M_2 is $r(2 - r)/3$. The odds of failing to diagnose A_2A_2 based on these marker genotypes are therefore $r(2 - r)/(3 - 2r + r^2)$, or approximately 2r/3 for small r.

Two kinds of error can evidently occur: incorrectly predicting the A_2A_2 genotype based on the *M* locus genotype, with probability $1 - (1 - r)^2$, or failing to predict the A_2A_2 genotype based on the *M* locus genotype, with probability $r(2 - r)/3$. In a situation like that of sickle cell anemia where antenatal diagnosis of a genetic pathology is desired, the costs associated with each type of error may be very different. We will not attempt to evaluate these relative costs, but will simply note that the probability of error goes to zero as r goes to zero. Clearly, very tightly linked loci are the most desirable, while even moderately linked loci may entail an unacceptably high error cost. In the discussion which follows we will restrict our attention to cases of very tight linkage, where the probability of recombination can be neglected.

An important aim in situations such as the sickle cell case is to determine whether or not a fetal offspring of a heterozygous carrier is A_2A_2, based on the *M* locus genotype. Two classes of mating are relevant: those between two heterozygotes (A_1A_2 by A_1A_2), and those between a heterozygote and an affected individual (A_1A_2 by A_2A_2). Not all of these matings will have the appropriate *M* locus markers for diagnosis. For instance, Table 1 shows that of the ten mating types which are A_1A_2 by A_1A_2, only those between double heterozygotes in the same linkage phase yield an exact diagnosis (assuming r = 0). In the other five matings in which at least one parent is heterozygous at the marker locus, the A_2A_2 condition can be ruled out half of the time. As a specific example, consider the mating A_1M_1 / A_2M_2 by A_1M_2 / A_2M_1 in which both parents are double heterozygotes, but in opposite linkage phases. Half of the pregnancies from such a couple will involve a fetus heterozygous at the marker locus, in which case no diagnosis is possible; the fetus could be either A_2A_2 or A_1A_1. In the other half of such pregnancies, the fetus would be homozygous at the marker locus and also A_1A_2. Hence, A_2A_2 can be ruled out in half of the pregnancies of this type of couple. The remaining three matings in which both parents are homozygous at the restriction site yield no information because all progeny will exhibit the same genotype at the marker locus.

The 12 matings involving A_1A_2 by A_2A_2 are presented in Table 2. In six of these matings the marker locus provides no information on the *A* locus genotype. Four matings yield an exact diagnosis, and in the remaining two cases diagnosis is possible half of the time.

Tables 1 and 2 assume that the linkage phase of the two loci is known in both parents. If the linkage phase is not apparent, then it must be determined indirectly through the testing of other offspring of the mating pair, or through the testing of parents or siblings of the mating pair. Knowledge of the linkage phase is essential to inferring progeny genotype based on associated marker loci in matings involving double heterozygotes. It is clear from Tables 1 and 2 that even when both parental two-locus genotypes are known, only a fraction of matings will permit a diagnosis.

Consequently, any reasonable evaluation of the utility of linked marker genes for predicting genetic transmission must consider the expected proportion of matings which are informative. This expected value is a function of the frequencies of the various gametic types in the population, as shown in the next section.

Table 1. Fraction of A_1A_2 by A_1A_2 matings in which the occurrence of A_2A_2 offspring can be confirmed or excluded on the basis of marker locus genotype.

Mating Type	Information	Frequency
A_1M_1/A_2M_1 by A_1M_1/A_2M_1	no information	$x_1^2x_3^2/p^2(1-p)^2$
A_1M_1/A_2M_1 by A_1M_1/A_2M_2	exclude A_2A_2 half the time	$2x_1^2x_3x_4/p^2(1-p)^2$
A_1M_1 A_2M_1 by A_1M_2/A_2M_1	exclude A_2A_2 half the time	$2x_1x_2x_3^2/p^2(1-p)^2$
A_1M_1/A_2M_1 by A_1M_2/A_2M_2	no information	$2x_1x_2x_3x_4/p^2(1-p)^2$
A_1M_1/A_2M_2 by A_1M_1/A_2M_2	exact diagnosis	$x_1^2x_4^2/p^2(1-p)^2$
A_1M_1/A_2M_2 by A_1M_2/A_2M_1	exclude A_2A_2 half the time	$2x_1x_2x_3x_4/p^2(1-p)^2$
A_1M_1/A_2M_2 by A_1M_2/A_2M_2	exclude A_2A_2 half the time	$2x_1x_2x_4^2/p^2(1-p)^2$
A_1M_2/A_2M_1 by A_1M_2/A_2M_1	exact diagnosis	$x_2^2x_3^2/p^2(1-p)^2$
A_1M_2/A_2M_1 by A_1M_2/A_2M_2	exclude A_2A_2 half the time	$2x_2^2x_3x_4/p^2(1-p)^2$
A_1M_2/A_2M_2 by A_1M_2/A_2M_2	no information	$x_2^2x_4^2/p^2(1-p)^2$

IV. POPULATION VARIABLES

To investigate how the expected fraction of informative A_1A_2 by A_1A_2 and A_1A_2 by A_2A_2 matings depends upon the population variables, it is necessary to specify the two-locus system precisely. Denote the frequencies of the four gametic types A_1M_1, A_1M_2, A_2M_1 and A_2M_2 by x_1, x_2, x_3 and x_4, respectively. We assume fitness

Table 2. Fraction of A_1A_2 by A_2A_2 matings in which the occurrence of A_2A_2 offspring can be confirmed or excluded on the basis of marker locus genotype.

Mating Type	Information	Frequency
A_1M_1/A_2M_1 by A_2M_1/A_2M_1	no information	$x_1x_3^3/p(1-p)^3$
A_1M_1/A_2M_1 by A_2M_1/A_2M_2	no information	$2x_1x_3^2x_4/p(1-p)^3$
A_1M_1/A_2M_1 by A_2M_2/A_2M_2	no information	$x_1x_3x_4^2/p(1-p)^3$
A_1M_1/A_2M_2 by A_2M_1/A_2M_1	exact diagnosis	$x_1x_3^2x_4/p(1-p)^3$
A_1M_1/A_2M_2 by A_2M_1/A_2M_2	exact diagnosis half the time	$2x_1x_3x_4^2/p(1-p)^3$
A_1M_1/A_2M_2 by A_2M_2/A_2M_2	exact diagnosis	$x_1x_4^3/p(1-p)^3$
A_1M_2/A_2M_1 by A_2M_1/A_2M_1	exact diagnosis	$x_2x_3^3/p(1-p)^3$
A_1M_2/A_2M_1 by A_2M_1/A_2M_2	exact diagnosis half the time	$2x_2x_3^2x_4/p(1-p)^3$
A_1M_2/A_2M_1 by A_2M_2/A_2M_2	exact diagnosis	$x_2x_3x_4^2/p(1-p)^3$
A_1M_2/A_2M_2 by A_2M_1/A_2M_1	no information	$x_2x_3^2x_4/p(1-p)^3$
A_1M_2/A_2M_2 by A_2M_1/A_2M_2	no information	$2x_2x_3x_4^2/p(1-p)^3$
A_1M_2/A_2M_2 by A_2M_2/A_2M_2	no information	$x_2x_4^3/p(1-p)^3$

coefficients of w_{11}, w_{12}, w_{22} for the A_1A_1, A_1A_2 and A_2A_2 genotypes, respectively. The *M* locus is taken to be neutral with respect to selection. Gene frequencies at the *A* and *M* loci are $p = x_1 + x_2$ for A_1 and $q = x_1 + x_3$ for M_1, respectively. Assuming random mating and that selection is completed before mating, the two-locus genotypic frequencies among mating adults are easily calculated. For example, the frequency of genotype A_1M_1/A_2M_1 is $2x_1x_3w_{12}/W$, where

$$W = p^2w_{11} + 2p(1 - p)w_{12} + (1 - p)^2 w_{22}$$

The same reasoning shows that the frequency of mating type A_1M_1/A_2M_1 by A_1M_1/A_2M_2 is $8x_1^2x_3x_4w_{12}^2/W^2$, and the total fraction of matings that are A_1A_2 by A_1A_2 is $4p^2(1-p)^2\ w_{12}^2/W^2$. The conditional probability of mating type A_1M_1/A_2M_1 by A_1M_1/A_2M_2 given A_1A_2 by A_1A_2 is therefore $2x_1^2x_3x_4/p^2(1-p)^2$. The other conditional mating frequencies are computed in a similar manner and are given in the last columns of Tables 1 and 2. Note that the same conditional frequency distributions apply if both loci are assumed to be selectively neutral.

From Table 1 the expected proportion of A_1A_2 by A_1A_2 matings in which the A_2A_2 condition can be diagnosed based on the *M* locus genotype is

$$\alpha_1 = (x_1x_4 + x_2x_3)/p(1 - p) - x_1x_2x_3x_4/p^2(1 - p)^2 \quad (1)$$

Similarly, Table 2 shows that the expected proportion of A_1A_2 by A_2A_2 matings in which the A_2A_2 condition can be ascertained is

$$\alpha_2 = (x_3^2 + x_4^2)/(1 - p)^2 - (x_1x_3^3 + x_2x_4^3)/p(1 - p)^3 \quad (2)$$

In analyzing these expressions, it is informative to express the gametic frequency distribution in terms of the marginal gene frequencies (p and q) and the linkage disequilibrium ($D = x_1x_4 - x_2x_3$) between the loci. This can be done via the transformation

$$x_1 = pq + D\ , \qquad x_2 = p(1 - q) - D$$
$$x_3 = (1 - p)q - D\ , \qquad x_4 = (1 - p)(1 - q) + D \quad (3)$$

where, for fixed values of p and q, D is restricted to the interval $-d_1 \leq D \leq d_2$ with

$$d_1 = \min[pq,\ (1-p)(1-q)] \quad \text{and} \quad d_2 = \min[p(1-q),\ (1-p)q]$$

Under this coordinate system, α_1 and α_2 become:

$$\begin{aligned}\alpha_1(p,q,D) &= q(1-q)(q^2-q+2) + D\{(1-2p)(1-2q)[1-q(1-q)]/p(1-p)\} \\ &+ D^2\{2[1-q(1-q)]/p(1-p)\} \\ &+ D^2\{[pq+(1-p)(1-q)][p(1-q)+q(1-p)]/p^2(1-p)^2\} \\ &- D^3[(1-2p)(1-2q)/p^2(1-p)^2] - D^4[1/p^2(1-p)^2] \qquad (4)\end{aligned}$$

and

$$\begin{aligned}\alpha_2(p,q,D) &= 2q(1-q)(q^2-q+1) \\ &+ D\{(1-2q)[2p+(1-4p)(q^2-q+1)]/p(1-p)\} \\ &+ D^2\{2p+3(1-2p)[q^2+(1-q)^2]\}/p(1-p)^2 \\ &+ D^3[(3-4p)(1-2q)/p(1-p)^3] + D^4[2/p(1-p)^3] \qquad (5)\end{aligned}$$

Both fractions depend upon the gene frequencies at each of the two loci as well as the linkage disequilibrium between them. It should also be noted that the explicit formulae (4) and (5) reveal a higher natural symmetry in the definition of α_1. In particular, the proportion of A_1A_2 by A_1A_2 matings which permit diagnosis (α_1) is symmetric in the gene frequencies at both loci, with

$$\alpha_1(p,q,D) = \alpha_1(1-p,q,-D) = \alpha_1(p,1-q,-D) = \alpha_1(1-p,1-q,D)$$

Hence, if the gene frequency at either locus is replaced by its complement, the range of D and the graph of α_1 as a function of D are simply reflected about D = 0. If both gene frequencies are replaced by their complements, the graph of α_1 is unchanged. In contrast, analysis of (5) shows that the fraction of informative A_1A_2 by A_2A_2 matings (α_2) is symmetric only in the gene frequency, q, at the marker locus.

The function α_1 was originally introduced and analyzed by Asmussen and Clegg (1982b). Despite its lesser symmetry, α_2, in many important respects, exhibits a qualitatively similar dependence upon the population variables. For example, when D = 0, $\alpha_2(q) = 2q(1-q)(q^2-q+1)$ ranges from a minimum of 0 (q = 0 or 1) to a maximum of 3/8 (q = 1/2), whereas $\alpha_1(q) = q(1-q)(q^2-q+2)$ ranges from 0 (q = 0 or 1) to 7/16 (q = 1/2). Hence, when the two loci are in linkage equilibrium (D = 0), the fetal genotype at the marker locus will allow diagnosis of the deleterious A_2A_2 condition in less than 44% of A_1A_2 by A_1A_2 pregnancies, and less than 38% of A_1A_2 by A_2A_2 pregnancies. Conversely, when there is absolute association between the two loci (*i.e.*, $x_1 + x_4 = 1$ or $x_2 + x_3 = 1$, and hence D at its maximum in absolute value), both α_1 and α_2 are one. These preliminary results are indicative of the dependence of α_1 and α_2 on the extent of linkage disequilibrium in the population. Indeed, Kan and Dozy (1978a) first proposed using the *Hpa* I polymorphism associated with the β-globin locus for genetic counseling after discovering that the 13 kbp fragment was in strong linkage disequilibrium with the β^S allele.

When p = q = 1/2, the influence of D on α_1 and α_2 can be easily analyzed. In this case, $\alpha_1(D) = 7/16 + 10D^2 - 16D^4$ ranges from 7/16 to 1 and $\alpha_2(D) = 3/8 + 8D^2 + 32D^4$ ranges from 3/8 to 1, as $|D|$ ranges from 0 to its maximum value of 1/4. Further analysis shows that whenever p = 1/2 or q = 1/2, α_1 is symmetric about D = 0 and increases monotonically as the linkage disequilibrium between the loci increases in magnitude. In contrast, α_2 is a monotonic function of $|D|$ only when the marker gene frequency q = 1/2. For other, nonsymmetric, gene frequencies, neither α_1 nor α_2 is minimized at D = 0, although numerical calculations show that the values of both functions at D = 0 are close to their minima (within 0.1).

Figures 2a,b,c illustrate the behavior of α_1 as a function of D for fixed values of p and q. Numerical examples indicate that if one or both loci are near fixation (and neither gene frequency is

a.

p=0.01
α_1
1.0
0.8
0.6
0.4
0.2
q=0.5
q=0.3
q=0.1
−0.012
−0.006
0.000
0.006
0.012
D

b.

p=0.2
α_1
1.0
0.8
0.6
0.4
0.2
q=0.5
q=0.3
q=0.1
−0.13
−0.06
0.00
0.06
0.13
D

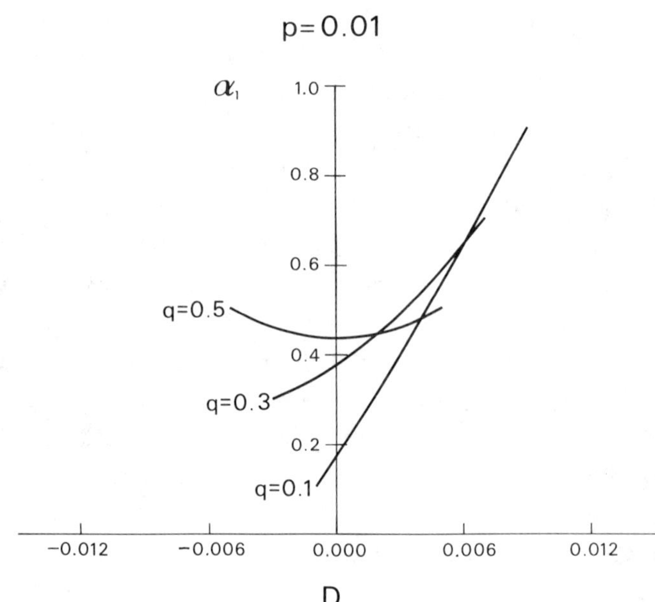

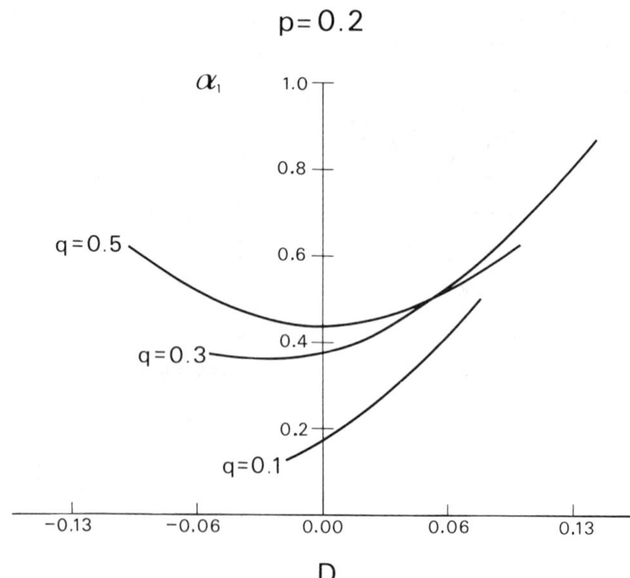

c.

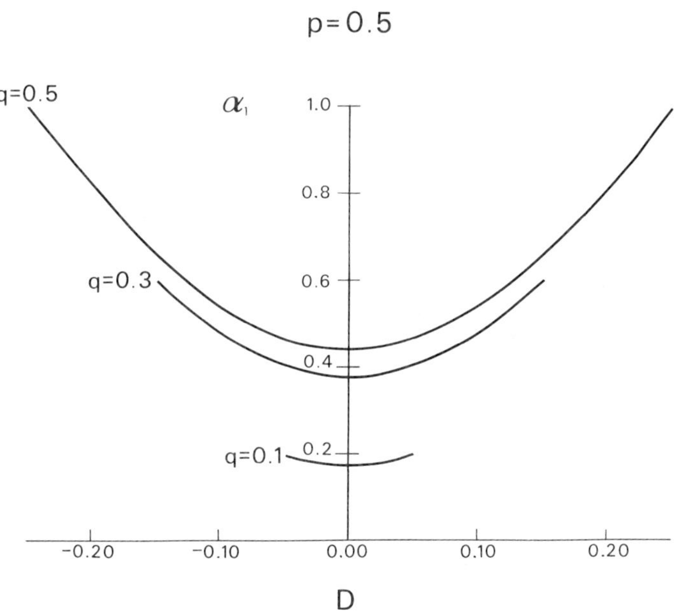

Figure 2. α_1 as a function of D, with q = 0.1, 0.3, 0.5, and (a) p = 0.01, (b) p = 0.20, (c) p = 0.50.

near 1/2), α_1 either strictly increases or strictly decreases across the entire range of admissible D values $[-d_1, d_2]$. Otherwise, α_1 has a minimum at a point $D = d^*$ within $(-d_1, d_2)$, increasing as D increases or decreases from d^*. From the sign of $\partial\alpha_1/\partial D$ at $D = 0$, α_1 will be minimized at a negative value of D in $[-d_1, 0)$ if both gene frequencies lie either above or below 1/2 [*i.e.*, $(p-1/2)(q-1/2) > 0$]. If, instead, one gene frequency lies above 1/2 and the other below 1/2, α_1 will be minimized at a positive value of D in $(0, d_2]$.

The graphs of α_2 exhibit the same range of behavior, as seen in Figures 3a,b,c,d. On the basis of the sign of $\partial\alpha_2/\partial D$ at $D = 0$, the location of the minimum of α_2 is instead governed by the sign of $(q - 1/2)[p - h(q)]$, where

a.

p=0.05
α_2
1.0
0.8
0.6
0.4
0.2
q=0.55
q=0.25
q=0.05
−0.05
−0.025
0.000
0.025
0.05
D

b.

p=0.25
α_2
1.0
0.8
0.6
0.4
0.2
q=0.55
q=0.25
q=0.05
−0.20
−0.10
0.00
0.10
0.20
D

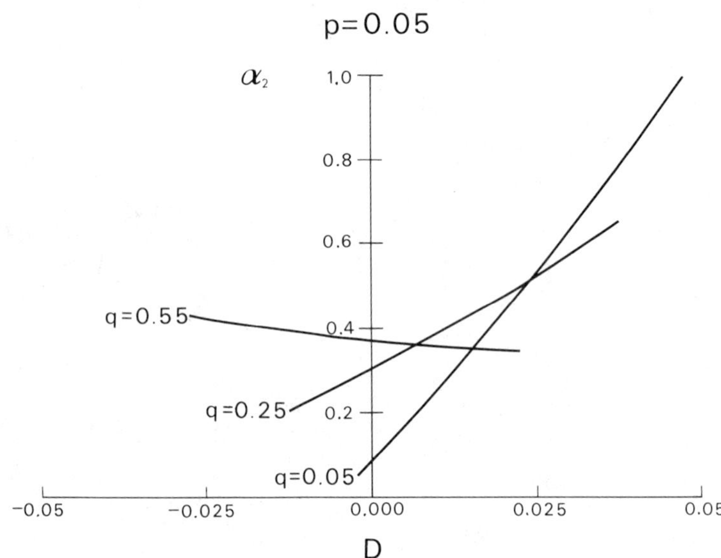

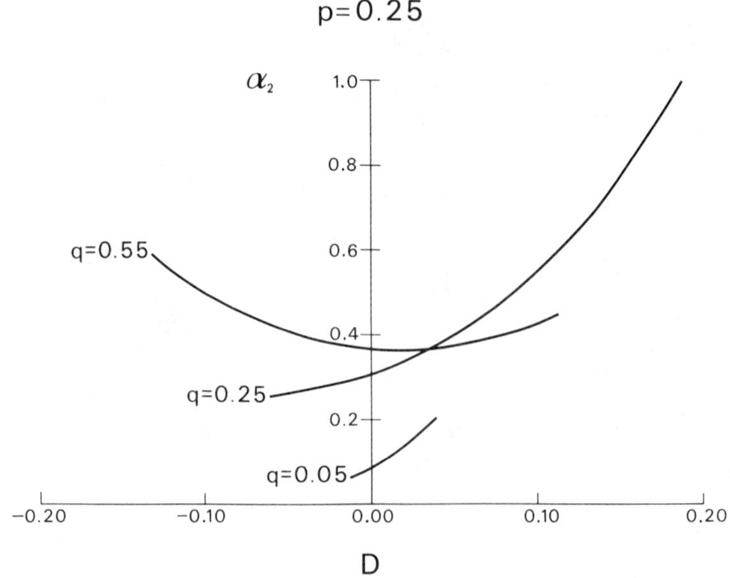

c.

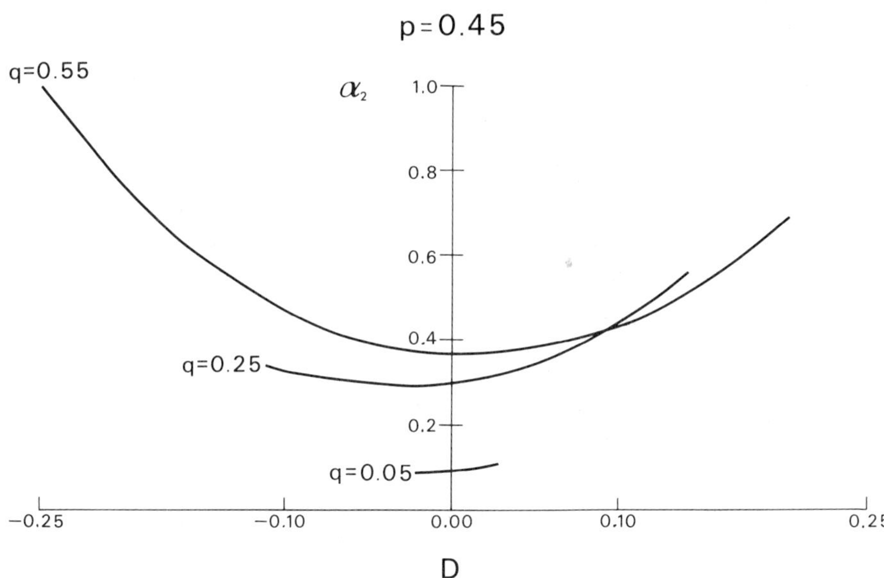

d.

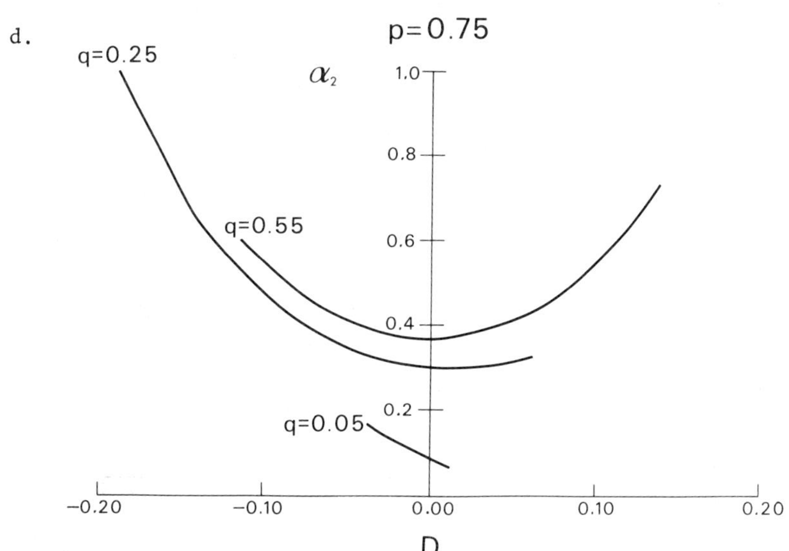

Figure 3. α_2 as a function of D, with q = 0.05, 0.25, 0.55, and (a) p = 0.05, (b) p = 0.25, (c) p = 0.45, (d) p = 0.75.

$$h(q) = \frac{(q - 1/2)^2 + 3/4}{4(q - 1/2)^2 + 1}$$

decreases from 3/4 to 1/2 as (q - 1/2) increases from 0 to 1/2.

An important discovery from numerical investigations is that the maximum values of α_1 and α_2 always occur when $D = -d_1$ or $D = d_2$, denoting complete association between the two loci (*i.e.*, exactly one gamete frequency x_i is 0). The corresponding values of α_1 and α_2 from (1) and (2) are given in Table 3. Note that the two functions agree when x_3 or x_4 is zero, but α_2 is less than or equal to α_1 when x_1 or x_2 is zero. Assuming that α_1 is in fact maximized at $D = -d_1$ or $D = d_2$, it can be proved from the values in Table 3 that α_1 is always maximized when D is at its maximum magnitude (Asmussen and Clegg, 1982b). Due to the asymmetry with respect to the gene frequency, p, at the selected locus, this is not precisely true of α_2. In particular, if p is near 1/2, the maximum of α_2 may occur at the endpoint (of admissible D values) with smaller magnitude. These are the only exceptions found, however, and the differences in these cases between the two endpoints $|-d_1| = d_1$ and d_2, and between the corresponding values of α_2 are always small. The intuitive notion that the most informative genetic markers are those which are in strong linkage disequilibrium with the target locus is therefore justified by the formal analysis.

Table 3. Expressions for α_1 and α_2 and corresponding values of D when there is complete association between loci.

Gamete	D	α_1	α_2
$x_1 = 0$	$-pq$	$q/(1-p)$	$q/(1-p) - q^2(1-p-q)/(1-p^3)$
$x_2 = 0$	$p(1-q)$	$(1-q)/(1-p)$	$(1-q)/(1-p) - (1-q)^2(q-p)/(1-p)^3$
$x_3 = 0$	$(1-p)q$	q/p	q/p
$x_4 = 0$	$-(1-p)(1-q)$	$(1-q)/p$	$(1-q)/p$

Some important observations about the dependence of α_1 on the two gene frequency variables can be obtained from an analysis of the maximum value of α_1, $\max(\alpha_1)$. Specifically, for any pair of gene frequencies at the two loci,

$$\max(\alpha_1) = \begin{cases} \dfrac{\min\ (q,\ 1-q)}{\min\ (p,\ 1-p)} & \text{if } \min(q,\ 1-q) \leq \min(p,\ 1-p) \\[2ex] \dfrac{\max\ (q,\ 1-q)}{\max\ (p,\ 1-p)} & \text{if } \min(q,\ 1-q) > \min(p,\ 1-p) \end{cases}$$

(Asmussen and Clegg, 1982b). $\text{Max}(\alpha_1)$ is thus the ratio of the smaller gene frequency at the marker locus to the smaller gene frequency at the selected locus, if the neutral marker locus is less polymorphic than the selected locus. Otherwise, $\max(\alpha_1)$ is the ratio of the two larger gene frequencies. $\text{Max}(\alpha_1)$ ranges from 0 to 1 in the first case, and from 1/2 to 1 in the second, depending on the exact values of the two gene frequencies. The maximum of α_1 is 1 if there is absolute association between the loci (*i.e.*, $x_2 = x_3 = 0$ which implies $p = q$ or $x_1 = x_4 = 0$ which implies $p = 1-q$). If the neutral locus is significantly less polymorphic than the selected locus [$\min(q,\ 1-q) \ll \min(p,\ 1-p)$], then α_1 will be small for all admissible values of D. This situation arises when q is near 0 or 1 and p is intermediate (see Figure 2c). On the other hand, α_1 will be near 1/2 for the entire range of admissible D values if q is intermediate and the selected locus is near fixation (see Figure 2a). More precisely, α_1 is bounded by 7/16 and 1/2 if $q = 1/2$ and $p = 0$ or 1. By continuity, when q is in the neighborhood of 1/2 and the frequency of the selected gene is near 0 or 1, α_1 will range from slightly below 7/16 to slightly above 1/2 as $|D|$ ranges from 0 to its maximum value. In these latter cases, little additional information is conveyed by the existence of linkage disequilibrium between the marker gene and the selected gene. The most informative situation arises when the gene frequencies at the two loci are similar (*i.e.*, $p \approx q$ or $p \approx 1 - q$) and $|D|$ is large in magnitude.

Due to the asymmetry in the definition of α_2, no simple formula for its maximum value is evident. Nonetheless, the dependence of α_2 on the two gene frequencies is qualitatively similar to that of α_1, although different in detail as Table 3 and Figure 3 show. An immediate observation is that both α_1 and α_2 are 0 if q is 0 or 1. Analysis of (5) indicates, moreover, that like α_1, α_2 is small for all admissible values of D if q is near 0 or 1 and p is intermediate. The degree of association between the loci also has little effect upon α_2 if the selected locus is near fixation and the marker locus is at an intermediate gene frequency. In particular, if p is near 0 and q is intermediate, α_2 ranges from slightly below to slightly above 3/8. A specific example is given by p = 0.05 and q = 0.45, for which α_2 ranges from approximately 0.35 to 0.44. If, instead, p is near 1 and q is near 1/2, α_2 ranges from slightly below 3/8 to slightly above 1/2. A final important similarity to α_1 is that α_2 is largest when the gene frequencies at the two loci are approximately equal and $|D|$ is large. Numerical examples indicate, however, that the marker locus is generally more informative in A_1A_2 by A_1A_2 matings than in A_1A_2 by A_2A_2 matings in that, for fixed values of p, q, and D, α_2 is usually less than α_1.

V. EVOLUTIONARY DYNAMICS

The utility of RFLPs as genetic markers rests upon exploiting naturally occurring genetic variants. We have seen that the expected proportion of matings in which the marker gene provides information on the transmission of the associated gene depends strongly on the two gene frequencies and the linkage disequilibrium. The values that these variables are likely to assume depend, in turn, upon the mode of selection, the rate of recombination and stochastic factors.

We will continue to assume that the marker locus is neutral with respect to selection and we will neglect the influence of stochastic factors such as finite population size. Subject to these assumptions, the recursion system for the gene frequencies and linkage disequilibrium is given by

$$p_{t+1} = p_t w_{1,t}/W_t \tag{6a}$$

$$q_{t+1} = q_t + D_t(w_{1,t} - w_{2,t})/W_t \tag{6b}$$

$$D_{t+1} = D_t(p_t) \tag{6c}$$

where

$$f(p_t) = (w_{1,t}\, w_{2,t} - r\, w_{12} W_t)/W_t^2 \tag{7}$$

and where

$$w_{1,t} = w_{12} + p_t(w_{11} - w_{12}),\ w_{2,t} = w_{22} + p_t(w_{12} - w_{22})$$

$$W_t = p_t w_{1,t} + (1 - p_t) w_{2,t}$$

and t is time measured in generations. Equations (6) and (7) represent the deterministic two-locus "hitchhiking" model which has been analyzed extensively by Thomson (1977) and Asmussen and Clegg (1981, 1982a).

A number of important features are evident from the transformation (6)-(7). First, the gene frequency at the selected locus follows the usual single locus equation (6a), independent of the other two variables. Hence p_t will converge monotonically to the polymorphic equilibrium

$$\hat{p} = \frac{w_{12} - w_{22}}{2w_{12} - w_{11} - w_{22}}$$

if there is overdominant selection (*i.e.*, $w_{12} > w_{11}, w_{22}$).

Otherwise, the frequency of one allele will steadily decline to zero, while the other allele monotonically approaches fixation. The rate of approach to equilibrium depends upon the mode and strength of selection. Second, the dynamics of the neutral process (6b) are governed by both the selected locus (including the gene frequency and form of selection) and the association between the two loci. In particular, q will change every generation that the two loci. In particular, q_t will change every generation that the is not at the polymorphic equilibrium $\hat{p}$ given above. Finally, the time dependent behavior of the linkage disequilibrium function under (6c) is determined by the gene frequency at the selected locus, the strength of selection and the degree of linkage between the two loci. Its qualitative behavior has been described by Asmussen and Clegg (1981) under various modes of selection.

An important discovery from the analysis of (6c) near the various gene frequency equilibria of the selected locus is that D_t always ultimately goes to zero as t becomes arbitrarily large (Asmussen and Clegg, 1982a). Consequently, the influence of the selected gene upon the neutral gene is transitory. The exact asymptotic rate of decay near fixation for A_i $(i = 1,2)$ is given by $(1 - r)w_{12}/w_{ii}$, and $1 - (rw_{12}/\hat{W})$ near the polymorphic equilibrium $\hat{p}$, where

$$\hat{W} = \frac{w_{12}^2 - w_{11}w_{22}}{2w_{12} - w_{11} - w_{22}}$$

Depending on the form and strength of selection, a significant amount of linkage disequilibrium may be generated between the two loci, however, before the ultimate decay process takes effect. The dynamical behavior of the association between the loci is now discussed for two particularly important cases.

A. Markers in Association with Genes Favored by Selection

Linkage disequilibrium is most apt to increase when a new favorable mutant enters a population at the target locus. For instance, the association between a neutral marker and a tightly linked ($r \leq 0.0001$) over-dominant mutant will initially increase rapidly and will continue to increase for virtually the entire trajectory of gene frequency change. The subsequent decay will usually occur on a much slower time scale, provided selection is strong.

It is again instructive to take the sickle cell case as a concrete example. A geographic survey of the RFLP associated with the β-globin locus suggests that, in West Africans, the $\beta^S(A_2)$ mutation arose on a 13.0 kbp fragment (M_1) at a restriction site which was already polymorphic in the population (Kan and Dozy, 1980). Under these conditions the two loci would have been in complete association initially with $D_0 = -(1 - p_0)(1 - q_0) < 0$. Approximate estimates of w_{11}, w_{12}, w_{22} are 0.9, 1.0, 0.3 (Cavalli-Sforza and Bodmer, 1971), yielding $\hat{p} = 0.875$. For $r = 0.00005$ (Kurnit and Hoehn, 1979), the linkage disequilibrium increases in magnitude while $0.8751 < p_t < 1$, and decays only once p_t is between $0.875 = \hat{p}$ and 0.8751. The rate of change is initially rapid, and slows as the equilibrium gene frequency is approached. For example, $f(p_t)$ in (6c) is 1.1099, 1.1074, 1.00002, 1.0, 0.99995 for p_t equal to 0.999, 0.9, 0.87511, 0.8751, 0.875, respectively. For a mutant with initial frequency $1 - p_0 = 0.01$, D_t/D_0 increases to roughly 11, whereas, if $1 - p_0 = 0.0001$, the initial linkage disequilibrium is increased over a hundred-fold before starting its eventual decline.

Gene frequency equilibrium and the associated increase in linkage disequilibrium is essentially achieved within 100 generations for the sickle cell case above. The decay of D_t, while accelerated at the equilibrium point, takes place on a much slower time scale. In fact, with $r = 0.00005$ and $\hat{W} = 0.913$, more than 12,000 generations are required for half of the linkage disequilibrium which existed when the gene frequency equilibrium was initially reached to decay. Thus the linkage disequilibrium generated by an episode of hitchhiking selection may persist for long periods when linkage is very tight. Under these circumstances we would expect α_1 to be closely approximated by $(1 - \hat{q})/\hat{p}$ which may remain close to 1 long after gene frequency equilibrium has been achieved, if q_0 was initially small (*i.e.*, if the initial state was close to one of absolute association).

More generally, Asmussen and Clegg (1981) proved that the linkage disequilibrium between a new favorable mutant (A_2) and a neutral marker locus will increase initially if $s > r$, where s is the selection intensity favoring heterozygotes for the new mutant. This applies to overdominant selection as well as all models of directional selection for which $w_{11} = 1 - s < 1 = w_{12}$. A weaker form of this result holds when the new favored mutant is recessive ($w_{11} = w_{12} = 1 - s < 1 = w_{22}$); there will be a transient increase in disequilibrium between the neutral and selected locus provided the selection intensity favoring the recessive phenotype is much greater than the linkage between the loci ($s >> r$). Whenever $s >> r$, very little recombination will take place during the course of gene frequency change so that the two loci will be near a state of complete association, and may remain close to this state for substantial periods of time. On these grounds, we expect most polymorphic genes strongly favored by selection to be in linkage disequilibrium with nearby restriction site polymorphisms. These considerations emphasize the importance of the initial gene frequency of the marker gene. If, for example, q_0 is initially high ($q_0 > 1/2$), then α_1 and α_2 will be less than 1/2. If, on the other hand, q_0 is small (less than 0.05), then α_1 and α_2 will be large and will remain large long after the selected gene has reached an equilibrium point.

B. Markers in Association with Rare Deleterious Genes

The case of rare deleterious genes is quite different. Asmussen and Clegg (1981) showed that $0 < f(p_t) < 1 - r$ when p_t is near 1, and A_2 is selected against ($w_{12} < w_{11}$ or $w_{22} < w_{12} = w_{11}$), so that D_t under (6c) decays, and at a rate faster than the neutral rate of $(1 - r)$ per generation. The exact size of the region, $(p^*, 1)$, where $f(p) < 1 - r$, depends on the mode of selection and on the values of the selection and recombination parameters, but this region is always moderately large. As a specific example, we consider a deleterious gene where the fitness coefficients w_{11}, w_{12}, w_{22} are 1, 1, 1 - s, respectively, with $0 < s \leq 1$. Then

$$f(p_t) = \frac{rsp_t^2 + s(1-2r)p_t + (1-r)(1-s)}{(1 - s + 2sp_t - sp_t^2)^2} \tag{8}$$

If $p_t > 3/4$, as would be the case when the deleterious gene is in low frequency, $0 < f(p_t) < 1 - r$ and increases to $1 - r$ as p_t increases to 1. In fact, decay will always be faster than the neutral rate $1 - r$ once $p_t \geq 1/2$.

Deleterious genes are usually thought to be maintained by a selection mutation balance. It would be desirable to use RFLPs for prenatal diagnosis of genetic diseases under such a system, and this type of application is technically feasible provided suitable cloned probes exist. To investigate this application, suppose A_1 mutates to A_2 at a rate μ per generation. Then, as shown by Asmussen and Clegg (1982b), the transformation (6) becomes

$$p_{t+1} = (1 - \mu)p_t\, w_{1,t}/W_t \tag{9a}$$

$$q_{t+1} = q_t + D_t\,(w_{1,t} - w_{2,t})/W_t \tag{9b}$$

$$D_{t+1} = D_t\, f(p_t)\,(1 - \mu) \tag{9c}$$

where $f(p_t)$ is defined in (7). If A_2 is again a deleterious recessive gene, then p_t will converge monotonically to its equilibrium value,

$$\hat{p} = 1 - (\mu/s)^{\frac{1}{2}}$$

provided $s > \mu$. Assuming the deleterious gene is rare, $f(p_t)$ will be initially less than $1 - r$, and if $s > 16\mu$, $f(p_t)$ will remain below 1 - r while converging monotonically to $f(\hat{p})$. Thus the linkage disequilibrium will decrease at a rate faster than $(1 - r)(1 - \mu)$ in every generation. The association between a neutral marker and a deleterious mutant will therefore decline even faster when the deleterious gene is maintained by mutation.

Returning to Figures 2 and 3, it is evident that q_0 must be low and $|D_0|$ high for α_1 or α_2 to be large when the new mutant is deleterious. That is, the initial state must be close to one of absolute association, which requires the somewhat unlikely event that a new mutant arises on a rare restriction fragment.

It should be emphasized that the transformation (9) applies regardless of the mode of selection, provided only that A_1 mutates to A_2 at a constant rate μ per generation. From (9c), mutation always decreases the per generation rate of change in D by the factor $1 - \mu$. In fact, recurrent mutation can also lead to much lower levels of association among RFLPs and selectively favored genes as is apparently the case when the global pattern of association between the β^S allele and the 13.0 kbp fragment is considered (Kan and Dozy, 1980).

VI. CONDITIONAL MEASURES OF ASSOCIATION

Exact diagnosis of the A_2A_2 condition from the fetal genotype at the marker locus in the procedure above requires that the parental two-locus linkage phases be known. If these are not available, conditional measures of association may be invoked to estimate the probability that the deleterious A_2A_2 condition is present. For instance, the probability that an individual received A_2 given that the marker allele M_1 was received is

$$\text{Prob}(A_2|M_1) = 1 - p - D/q \qquad (10a)$$

Similarly, the probability that an individual received A_2 if M_2 was received is

$$\text{Prob}(A_2|M_2) = 1 - p + D/(1 - q) \tag{10b}$$

For the marker locus to be worthwhile as a predictor of the deleterious A_2 gene, at least one of these conditional probabilities must be high. Like α_1 and α_2, they depend upon the gene frequencies at the two loci as well as on the linkage disequilibrium between them. When $D = 0$, both conditional probabilities in (10) equal $1 - p$, the frequency of the deleterious allele A_2, which will usually be small. If $D < 0$, $\text{Prob}(A_2|M_2) < 1 - p < \text{Prob}(A_2|M_1)$, so that M_1 is the only feasible predictor of A_2. From (10a), $\text{Prob}(A_2|M_1)$ decreases linearly with D; it is maximized when D is at its maximal negative magnitude $- d_1$, where

$$\max \text{Prob}(A_2|M_1) = \begin{cases} 1 & \text{if } q \leq 1 - p \\ \dfrac{(1 - p)}{q} & \text{if } q > 1 - p \end{cases}$$

Analogous results hold for the marker allele M_2 when $D > 0$. The marker locus is therefore an effective predictor of A_2 only if the deleterious allele A_2 is nearly as frequent as (or more frequent than) the associated marker allele M_1 (M_2) and D is nearly maximally negative (positive).

Although it is difficult to analyze the dynamical behavior of (10) when the target locus is selected, certain approximate statements can be made (Asmussen and Clegg, 1982b). Suppose, as in the sickle cell case, that the A_2 mutation occurs in a population polymorphic at the marker locus, and that A_2 is initially associated with the marker allele M_1. Under these conditions, the linkage disequilibrium between the loci is initially negative and will remain negative under the hitchhiking models (6) and (9). Consider first the case where A_2 is a recessive deleterious gene with $w_{22} < w_{11} = w_{12}$. Then its frequency, $1 - p_t$, would ordinarily be small initially and remain small. Moreover, q_t under (6b) and (9b) will change only slightly, so that p_t and q_t can be roughly approximated by their initial values p_0 and q_0. Finally, as indicated above, the linkage disequilibrium declines in magnitude each generation. Under these conditions, an approximate upper bound for the conditional probability (10a) is given by

$$\mathrm{Prob}(A_2|M_1) \leq \frac{(1 - p_0)}{q_0}$$

This quantity clearly declines as q_0, the initial frequency of the associated marker allele, increases. In fact, numerical iterations of the recursion system (9) show that the exact conditional probability (10a) is usually small, as illustrated in Table 4 for the case of a recessive lethal.

If the new mutant is instead overdominant, the two gene frequencies may change substantially and, as shown above, D may increase in magnitude for virtually the entire trajectory of gene

Table 4. Values of the conditional association $\mathrm{Prob}(A_2|M_1)$ after 200 generations as a function of q_0, the initial frequency of marker allele M_1 associated with the deleterious mutant A_2.

	Recessive Lethal Selection		Overdominant Selection	
q_0	$p_0 = 0.999$	$p_0 = 0.99$	$p_0 = 0.999$	$p_0 = 0.99$
0.002	0.410	-	0.986	-
0.02	0.037	0.228	0.876	0.926
0.1	0.009	0.033	0.586	0.606
0.3	0.004	0.011	0.321	0.326
0.5	0.003	0.007	0.221	0.223
0.9	0.002	0.005	0.137	0.137

frequency change. In numerical iterations of (6) with parameters approximating the sickle cell case, $\text{Prob}(A_2|M_1)$ is large only for small values of q_0, as seen in Table 4. In this table, fitness parameters are $w_{11} = w_{12} = 1$, $w_{22} = 0$ for the recessive lethal model, and $w_{11} = 0.9$, $w_{12} = 1$, $w_{22} = 0.3$ for the overdominant model. In both cases the recombination parameter is $r = 5 \times 10^{-5}$ and the initial frequency of the deleterious mutant A_2 is $1 - p_0$. The mutation parameter in the recessive lethal model is $\mu = 10^{-5}$. These calculations further highlight the way in which the initial frequency of the marker gene determines its value as a predictor of genetic transmission.

VII. CONCLUSIONS

Two major applications of marker genes are (i) the mapping and genetic dissection of major phenotypic traits, and (ii) the predicting of transmission of other genes in statistical association with the marker gene. The requirements of these two applications differ in several important respects. Botstein *et al.* (1980) estimate that approximately 150 polymorphic restriction sites distributed about 20 cM (centiMorgans) apart would be sufficient to construct a detailed linkage map of the human genome (number modified in the chapter by Bishop *et al.* in this volume). Once such an inventory of genetic markers existed, they could routinely be employed in mapping other genetic traits. Relatively weak to moderate linkage ($0.35 \geq r \geq 0.10$) would be quite adequate for this purpose, and traditional methods of genetic analysis, *i.e.*, the estimation of recombination fractions from segregating families or pedigrees, could be used. Once a gene has been mapped, there is little call for repetition of the task.

In contrast to the mapping application, the use of marker genes to predict the transmission of associated genes emphasizes repeatability. The larger the proportion of families in which the marker gene is informative, the greater the utility of the technique. The proportion of informative matings depends in turn on the gene frequency and linkage disequilibrium variables which are themselves controlled by recombination and selection processes. Furthermore, unlike many population genetic models, initial conditions play a crucial role. Despite these complexities, certain general conclusions are warranted. First, the technique is most useful when linkage between the marker locus and target gene is very tight, as would occur when the target gene is physically located on the polymorphic restriction fragments. This represents the optimal case because (i) the error cost is minimized, and (ii) high levels of association between the marker locus and the target gene are expected to persist for long periods of time. Second, overdominant selection or selection favoring rare genes can lead to a substantial increase in linkage disequilibrium, while selection against rare deleterious genes is associated with a decay in linkage disequilibrium, even when maintained by recurrent mutation. Hence, the use of markers to predict the transmission of rare deleterious genes is likely to have less application than in the overdominant case illustrated by sickle cell anemia. Third, recurrent mutation also causes a decay in linkage disequilibrium and may be quite significant when the global distribution of a polymorphism is considered. Fourth, the optimal initial condition is one of absolute association. When this holds, all matings will initially be informative, and with r near zero, this condition will persist for many generations.

Finally, the proportion of informative matings depends on the number of alleles at the marker locus. (Tightly linked polymorphic sites can be regarded as multiple alleles at a single locus.) Preliminary calculations for the A_1A_2 x A_1A_2 mating type show that α_1 for the triallelic case is at least as great as, and usually greater than α_1 for the diallelic case, discussed in this chapter (Asmussen and Clegg, unpublished results). Moreover, the fraction of informative matings continues to increase as the number of alleles at the marker locus increases beyond three.

At the present time, the major application of RFLPs in predicting genetic transmission has been in medical genetics and in genetic counseling. Because the DNA techniques are not dependent on gene expression and can therefore be applied at very early stages of development, RFLPs are potentially valuable in predicting the transmission of any gene whose expression occurs late in development. The primary limiting factor in the application of RFLPs as predictors of genetic transmission is the availability of cloned DNA sequences of the target gene. Appropriate cloned sequences do exist for globin genes (Kurnit and Hoehn, 1979) and the insulin gene (Rotwein *et al*., 1981) in man. If current estimates of the level of DNA sequence polymorphism are correct, then it is likely that appropriate RFLPs associated with any cloned gene of interest can be detected due to the large variety of restriction endonucleases presently available. We expect that, as cloned genes accumulate, the use of RFLPs as markers will have an increasing application.

ACKNOWLEDGMENTS

This work was supported by National Science Foundation grants DEB81-18414 and DEB82-00664. This chapter was prepared while M. T. Clegg held a John Simon Guggenheim Foundation Fellowship and was a Visiting Scientist at the CSIRO Division of Plant Industry.

BIBLIOGRAPHY

Adams, J., and E. D. Rothman (1982). Estimation of phylogenetic relationships from DNA restriction patterns and selection of endonuclease cleavage sites. Proc. Natl. Acad. Sci. USA 79 : 3560-3564.

Anderson, S., A. T. Bankier, B. G. Barrell, M. H. L. de Bruijn, A. R. Coulson, J. Drouin, I. C. Eperon, D. P. Nierlich, B. A. Roe, F. Sanger, P. H. Schreier, A. J. H. Smith, R. Staden and I. G. Young (1981). Sequence and organization of the human mitochondrial genome. Nature 290 : 457-465. (Erratum 291 168.)

Anderson, S., M. H. L. de Bruijn, A. R. Coulson, I. C. Eperon, F. Sanger and I. G. Young (1982). Complete sequence of bovine mitochondrial DNA. Conserved features of the mammalian mitochondrial genome. J. Mol. Biol. (to appear)

Andrews, P. (1982). Hominoid evolution. Nature 295 : 185-186.

Aoki, K., Y. Tateno and N. Takahata (1981). Estimating evolutionary distance from restriction maps of mitochondrial DNA with arbitrary G+C content. J. Mol. Evol. 18 : 1-8.

Aquadro, C. F., and B. D. Greenberg (1983). Human mitochondrial DNA sequence variation and evolution. Genetics : 287-312.

Arnberg, A. C., G. J. B. Van Ommen, L. A. Grivell, E. F. J. Van Bruggen and P. Borst (1980). Some yeast mitochondrial RNAs are circular. Cell 19 : 313-319.

Arnheim, N., M. Krystal, R. Schmickel, G. Wilson, O. Ryder and E. Zimmer (1980). Molecular evidence for genetic exchanges among ribosomal genes on nonhomologous chromosomes in man and apes. Proc. Natl. Acad. Sci. USA 77 : 7323-7327.

Asmussen, M. A., and M. T. Clegg (1981). Dynamics of the linkage disequilibrium function under models of gene-frequency hitchhiking. Genetics 99 : 337-356.

Asmussen, M. A., and M. T. Clegg (1982a). Rates of decay of linkage disequilibrium under two-locus models of selection. J. Math. Biology 14 : 37-70.

Asmussen, M. A., and M. T. Clegg (1982b). Use of restriction fragment length polymorphisms for genetic counselling: population genetic considerations. Am. J. Human Genet. 34 :369-380.

Auron, P. E., W. P. Rindone, C. P. H. Vary, J. J. Celentano and J. N. Vournakis (1982). Computer-aided prediction of RNA secondary structures. Nucleic Acids Res. 10 : 403-419.

Avise, J. C., C. Giblin-Davidson, J. Laerm, J. C. Patton and R. A. Lansman (1979). Mitochondrial DNA clones and matriarchal phylogeny within and among geographic populations of the pocket gopher, *Geomys pinetis*. Proc. Natl. Acad. Sci. USA 76 : 6694-6698.

Bach, R., P. Friedland, D. L. Brutlag and L. Kedes (1982). MAXAMIZE. A DNA sequencing strategy advisor. Nucleic Acids Res. 10 : 295-304.

Barker, W. C., and M. O. Dayhoff (1972). Detecting distant relationships: computer methods and results. IN Atlas of Protein Sequence and Structure. Dayhoff, M. O. (Ed.), Volume 6, National Biomedical Research Foundation, Silver Spring, Maryland.

Berget, S. M., C. Moore and P. A. Sharp (1977). Spliced segments at the 5' terminus of adenovirus 2 late mRNA. Proc. Natl. Acad. Sci. USA 74 : 3171-3175.

Bibb, M. J., R. A. Van Etten, C. T. Wright, M. W. Walberg and D. A. Clayton (1981). Sequence and gene organization of mouse mitochondrial DNA. Cell 26 : 167-180.

Bina, M., R. J. Feldmann and R. G. Deeley (1980). Could poly(A) align the splicing sites of messenger RNA precursors? Proc. Natl. Acad. Sci. USA 77 : 1278-1282.

Bishop, D. T., and M. H. Skolnick (1980). Numerical considerations for linkage studies using polymorphic DNA markers in humans. IN Banbury Report 4: Cancer Incidence in Defined Populations. Cairns, J., J. L. Lyon and M. Skolnick (Eds.), Cold Spring Harbor Laboratory, New York, pp. 421-433.

Bishop, D. T., and M. H. Skolnick (1983). Genetic markers and linkage studies. IN Banbury Report 14: Recombinant DNA Applications to Human Disease. Caskey, T., and R. White (Eds.), Cold Spring Harbor Laboratory, New York (to appear).

Bishop, Y. M. M., S. E. Fienberg and P. W. Holland (1975). Discrete Multivariate Analysis. MIT Press, Cambridge, Mass.

Blumenthal, R. M., P. J. Rice and R. J. Roberts (1982). Computer programs for nucleic acid sequence manipulation. Nucleic Acids Res. 10 : 91-101.

Bolivar, F., and K. Backman (1979). Plasmids of *Escherichia coli* as cloning vectors. IN Methods in Enzymology 68: Recombinant DNA. Wu, R. (Ed.), Academic Press, New York, pp. 245-267.

Bosman, F. T., M. van der Ploeg, P. von Duijn and A. Schaberg (1977). Photometric determination of the DNA distribution in the 24 human chromosomes. Exp. Cell Res. 105 : 301-311.

Bostian, K. A., R. C. Lee and H. O. Halvorson (1979). Preparative fractionation of nucleic acids by agarose gel electrophoresis. Anal. Biochem. 95 : 174-182.

Botstein, D., R. L. White, M. Skolnick and R. W. Davis (1980). Construction of a genetic linkage map in man using restriction fragment length polymorphisms. Am. J. Human Genet. 32 : 314-331.

Brown, A. H. D., M. W. Feldman and E. Nevo (1980). Multilocus structure of natural populations of *Hordeum spontaneum*. Genetics 96 : 523-536.

Brown, W. M. (1980). Polymorphism in mitochondrial DNA of humans as revealed by restriction endonuclease analysis. Proc. Natl. Acad. Sci. USA 77 : 3605-3609.

Brown, W. M., M. George and A. C. Wilson (1979). Rapid evolution of animal mitochondrial DNA. Proc. Natl. Acad. Sci. USA 76 : 1967-1971.

Brown, W. M., E. M. Prager, A. Wang and A. C. Wilson (1982). Mitochondrial DNA sequences of primates: tempo and mode of evolution. J. Mol. Evol. 18 : 225-239.

Brown, W. M., and J. Vinograd (1974). Restriction endonuclease cleavage maps of animal mitochondrial DNAs. Proc. Natl. Acad. Sci. USA 71 : 4617-4621.

Brutlag, D. L., J. Clayton, P. Friedland and L. H. Kedes (1982). SEQ: a nucleotide sequence analysis and recombination system. Nucleic Acids Res. 10 : 279-294.

Cameron, J. R., E. Y. Loh and R. W. Davis (1979). Evidence for transposition of dispersed repetitive DNA families in yeast. Cell 16 : 739-751.

Camin, J. H., and R. R. Sokal (1965). A method for deducing branching sequences in phylogeny. Evolution 19 : 311-326.

Cavalli-Sforza, L. L., and W. F. Bodmer (1971). The Genetics of Human Populations. Freeman, San Francisco.

Cavalli-Sforza, L. L., and A. W. F. Edwards (1967). Phylogenetic analysis: models and estimation procedures. Evolution 21 : 550-570. (Also Am. J. Human Genet. 19 : 233-257.

Cavender, J. A. (1978). Taxonomy with confidence. Math. Biosciences 40 : 271-280. (Erratum 44 : 309, 1979.)

Cavender, J. A. (1981). Tests of phylogenetic hypotheses under generalized models. Math. Biosciences 54 : 217-229.

Chakraborty, R. (1977). Estimation of time of divergence from phylogenetic studies. Can. J. Genet. Cytol. 19 : 217-223.

Chow, L. T., R. E. Gelinas, T. R. Broker and R. J. Roberts (1977). An amazing sequence arrangement at the 5' ends of adenovirus 2 messenger RNA. Cell 12 : 1-8.

Clayton, J., and L. Kedes (1982). GEL, a DNA sequencing project management system. Nucleic Acids Res. 10 : 305-321.

Clegg, M. T. (1982). DNA sequences in biosystematics and evolution. IN New Approaches to Biosystematics and Evolution. O. Solbrig (Ed.) (to appear).

Cooke, P. J. (1974). Bounds for coverage probabilities with applications to sequential coverage problems. J. Appl. Prob. 11 : 281-293.

Corruccini, R. S., M. Baba, M. Goodman, R. L. Ciochon and J. E. Cronin (1980). Non-linear macromolecular evolution and the molecular clock. Evolution 34 : 1216-1219.

Daniels, D. L., J. R. de Wet and F. R. Blattner (1980). New map of bacteriophage *lambda* DNA. J. Virology 33 : 390-400.

Davies, K. E., B. D. Young, R. G. Elles, M. E. Hill and R. Williamson (1981). Cloning of a representative genomic library of the human X chromosome after sorting by flow cytometry. Nature 293 : 374-376.

Davis, R. W., D. Botstein and J. R. Roth (1980). Advanced Bacterial Genetics. Cold Spring Harbor Laboratories, Cold Spring Harbor, New York.

Dempster, A. P., N. M. Laird and D. B. Rubin (1977). Maximum likelihood from incomplete data via the EM algorithm. J. R. Statist. Soc. B 39 : 1-22.

Dennis, E. S., and W. J. Peacock (1982). Knob heterochromatin and the evolution of maize. Science (submitted).

Douthwaite, S., A. Christensen and R. A. Garrett (1982). Binding sites of ribosomal proteins on prokaryotic 5S ribonucleic acids: a study with ribonucleases. Biochemistry 21 : 2313-2320.

Douthwaite, S., and R. A. Garrett (1981). Secondary structure of prokaryotic 5S ribosomal ribonucleic acids: a study with ribonucleases. Biochemistry 20 : 7301-7307.

Duggelby, R. G., H. Kinns and J. I. Rood (1981). A computer program for determining the size of DNA restriction fragments. Anal. Biochem. 110 : 49-55.

Dumas, J-P., and J. Ninio (1982). Efficient algorithms for folding and comparing nucleic acid sequences. Nucleic Acids Res. 10 : 197-216.

Eck, R. V., and M. O. Dayhoff (1966). Atlas of Protein Sequence and Structure. National Biomedical Research Foundation, Silver Spring, Maryland.

Edwards, A. W. F., and L. L. Cavalli-Sforza (1963). The reconstruction of evolution. Ann. Hum. Genet. 27 : 104. (Also Heredity 18 : 553.)

Edwards, A. W. F., and L. L. Cavalii-Sforza (1964). Reconstruction of evolutionary trees. IN Phenetic and Phylogenetic Classification. Systematics Association Publication No. 6, Heywood, V. H., and J. McNeill (Eds.), Systematics Association, London, pp. 67-76.

Elston, R. C., and K. Lange (1975). The prior probability of autosomal linkage. Ann. Hum. Genet. 38 : 341-350.

Engels, W. R. (1981). Estimating genetic divergence and genetic variability with restriction endonucleases. Proc. Natl. Acad. Sci. USA 78 : 6329-6333.

Estabrook, G. F., C. S. Johnson, Jr., and F. R. McMorris (1976). A mathematical foundation for the analysis of cladistic character compatibility. Math. Biosci. 29 : 181-187.

Ewens, W. J. (1972). The sampling theory of selectively neutral alleles. Theor. Pop. Biol. 3 : 87-112.

Ewens, W. J., R. S. Spielman and H. Harris (1981). Estimation of genetic variation at the DNA level from restriction endonuclease data. _Proc. Natl. Acad. Sci. USA_ _78_ : 3748-3750.

Fangman, W. L. (1978). Separation of very large DNA molecules by gel electrophoresis. _Nucleic Acids Res._ _5_ : 653-665.

Farris, J. S. (1973). A probability model for inferring evolutionary trees. _Syst. Zool._ _22_ : 250-256.

Farris, J. S. (1981). Distance in phylogenetic analysis. IN _Advances in Cladistics_. Funk, V. A., and D. R. Brooks (Eds.), New York Botanical Garden, New York, pp. 3-23.

Feldmann, R. J., D. H. Bing, B. C. Furie and B. Furie (1978). Interactive computer surface graphics approach to study of the active site bovine trypsin. _Proc. Natl. Acad. Sci. USA_ _75_ : 5409-5412.

Feller, W. (1968). _An Introduction to Probability Theory and Its Applications_. _Volume 1, Third Edition, John Wiley, New York._

Felsenstein, J. (1973). Maximum likelihood and minimum-steps methods for estimating evolutionary trees from data on discrete characters. _Syst. Zool._ _22_ : 240-249.

Felsenstein, J. (1978a). The number of evolutionary trees. _Syst. Zool._ _27_ : 27-33.

Felsenstein, J. (1978b). Cases in which parsimony or compatibility methods will be positively misleading. _Syst. Zool._ _27_ : 401-410.

Felsenstein, J. (1979). Alternative methods of phylogenetic inference and their interrelationship. _Syst. Zool._ _28_ : 49-62.

Felsenstein, J. (1981a). Evolutionary trees from DNA sequences: a maximum likelihood approach. _J. Mol. Evol._ _17_ : 368-376.

Felsenstein, J. (1981b). A likelihood approach to character weighting and what it tells us about parsimony and compatibility. _Biol. J. Linn. Soc._ _16_ : 183-196.

Felsenstein, J. (1983). The statistical approach to inferring evolutionary trees and what it tells us about parsimony and compatibility. IN _The Estimation of Evolutionary History: Proceedings of a Workshop on the Theory and Application of Cladistic Methodology_. Duncan, T., and T. F. Stuessy (Eds.), Columbia University Press, New York (in press).

Ferris, S. D., A. C. Wilson and W. M. Brown (1981). Evolutionary tree for apes and humans based on cleavage maps of mitochondrial DNA. _Proc. Natl. Acad. Sci. USA_ _78_ : 2432-2436.

Fitch, W. M. (1966). The relation between frequencies of amino acids and ordered trinucleotides. J. Mol. Biol. 16 : 1-8.

Fitch, W. M. (1971). Toward defining the course of evolution: minimum change for a specified tree topology. Syst. Zool. 20 : 406-416.

Fitch, W. M. (1975). Toward finding the tree of maximum parsimony. IN Proceedings of the Eighth International Conference on Numerical Taxonomy. Estabrook, G. F. (Ed.), Freeman, San Francisco, pp. 189-230.

Fitch, W. M. (1977). On the problem of discovering the most parsinomious tree. Amer. Natur. 111 : 223-257.

Fitch, W. M., and J. S. Farris (1974). Evolutionary trees with minimum nucleotide replacements from amino acid sequences. J. Mol. Evol. 3 : 263-278.

Fitch, W. M., and E. Margoliash (1967). Construction of phylogenetic trees. Science 155 : 279-284.

Flatto, L., and A. G. Konheim (1962). The random division of an interval and the random covering of a circle. SIAM Review 4 : 211-222.

Foulds, L. R., and R. L. Graham (1982). The Steiner problem in phylogeny is NP-complete. Adv. Appl. Math. 3 : 43-49.

Foulds, L. R., M. D. Hendy and D. Penny (1979). A graph theoretic approach to the development of minimal phylogenetic trees. J. Mol. Evol. 13 : 127-149.

Friedland, P., L. Kedes, D. Brutlag, Y. Iwasaki and R. Bach (1982). GENESIS, a knowledge-based genetic engineering simulation system for representation of genetic data and experiment planning. Nucleic Acids Res. 10 : 323-340.

Fuchs, C., E. C. Rosenvold, A. Honigman and W. Szybalski (1978). A simple method for identifying the palindromic sequences recognized by restriction endonucleases: the nucleotide sequence of the Ava II site. Gene 4 : 1-23.

Gillespie, J. H., and C. H. Langley (1979). Are evolutionary rates really variable? J. Mol. Evol. 13 : 27-34.

Gillham, N. W. (1978). Organelle Heredity. Raven Press, New York.

Gingeras, T. R., J. P. Milazzo and R. J. Roberts (1978). A computer assisted method for the determination of restriction enzyme recognition sites. Nucleic Acids Res. 5 : 4105-4127.

Gingeras, T. R., J. P. Milazzo, D. Sciaky and R. J. Roberts (1979). Computer programs for the assembly of DNA sequences. Nucleic Acids Res. 7 : 529-545.

Gingeras, T. R., P. Rice and R. J. Roberts (1982). A semi-automated method for the reading of nucleic acid sequencing gels. Nucleic Acids Res. 10 : 103-114.

Gingeras, T. R., and R. J. Roberts (1980). Steps toward computer analysis of nucleotide sequences. Science 209 : 1322 - 1328.

Gingeras, T. R., D. Sciaky, R. E. Gelinas, B. Jiang, C. E. Yen, M. M. Kelly, P. A. Bullock, B. L. Parsons, K. E. O'Neill and R. J. Roberts (1982). Nucleotide sequences from the Adenovirus-2 genome. J. Biol. Chem. 257 : 13475-13491.

Goad, W. B., and M. I. Kanehisa (1982). Pattern recognition in nucleic acid sequences. I. A general method for finding local homologies and symmetries. Nucleic Acids Res. 10 : 247-263.

Goodman, H. M., M. V. Olson and B. D. Hall (1977). Nucleotide sequence of a mutant eukaryotic gene: the yeast tyrosine-inserting ochre suppressor SUP4-o. Proc. Natl. Acad. Sci. USA 74 : 5453-5457.

Goodman, M. (1981). Globin evolution was apparently very rapid in early vertebrates: a reasonable case against the rate-constancy hypothesis. J. Mol. Evol. 17 : 114-120.

Gotoh, O., J-T. Hayashi, H. Yonekawa and Y. Tagashira (1979). An improved method for estimating sequence divergence between related DNAs from changes in restriction endonuclease cleavage sites. J. Mol. Evol. 14 : 301-310.

Grabowski, P. J., A. J. Zaug and T. R. Cech (1981). The intervening sequence of the ribosomal RNA precursor is converted to a circular RNA in isolated nuclei of tetrahymena. Cell 23 : 467-476.

Grantham, R., C. Gautier, M. Gouy, R. Mercier and A. Pave (1980). Codon catalog usuage and the genome hypothesis. Nucleic Acids Res. 8 : r49-r62.

Gusella, J. F., C. Keys, A. Varsanyi-Breiner, F-T. Kao, C. Jones, T. T. Puck and D. Housman (1980). Isolation and localization of DNA segments from specific human chromosomes. Proc. Natl. Acad. Sci. USA 77 : 2829-2833.

Haldane, J. B. S. (1919). The combination of linkage values and the calculation of distances between the loci of linked factors. J. Genetics 8 : 299-309.

Hamrick, J. L., Y. B. Linhart and J. B. Mitton (1979). Relationships between life history characteristics and electrophoretically detectable genetic variation in plants. Ann. Rev. Ecol. Syst. 10 : 173-200.

Hartigan, J. A. (1973). Minimum mutation fits to a given tree. Biometrics 29 : 53-65.

Hendy, M. D., D. Penny and L. R. Foulds (1978). Identification of phylogenetic trees of minimal length. J. Theor. Biol. 17 : 441-452.

Hudson, R. R. (1982). Estimating genetic variability with restriction endonucleases. Genetics 100 : 711-719.

Hulten, M. A., D. Laurie and R. Palmer (1981). Estimation of recombination fractions and genetic lengths from chiasma data. Sixth International Congress of Human Genetics, Jerusalem, Israel, 1981. Abstract.

Jacobs, L. L., and D. Pilbeam (1980). Of mice and men: fossil-based divergence dates and molecular "clocks". J. Human Evol. 9 : 551-555.

Jeffreys, A. J. (1979). DNA sequence variants in the G_{γ}-, G_{γ}-, δ- and β-globin genes of man. Cell 18 : 1-10.

Johnson, P. H., and L. I. Grossman (1977). Electrophoresis of DNA in agarose gels. Optimizing separations of conformational isomers of double- and single-stranded DNAs. Biochemistry 16 : 4217-4225.

Johnson, P. H., M. J. Miller and L. I. Grossman (1980). Electrophoresis of DNA in agarose gels. II. Effects of loading mass and electroendosmosis on electrophoretic mobilities. Anal. Biochem. 102 : 159-162.

Jukes, T. H., and C. R. Cantor (1969). Evolution of protein molecules. IN Mammalian Protein Metabolism, III. Munro, H. N. (Ed.), Academic Press, New York, pp. 21-132.

Kan, Y. W., and A. M. Dozy (1978a). Polymorphism of DNA sequence adjacent to human β-globin structural gene: relationship to sickle mutation. Proc. Natl. Acad. Sci. USA 75 : 5631-5635.

Kan, Y. W., and A. M. Dozy (1978b). Antenatal diagnosis of sickle-cell anaemia by DNA analysis of amniotic-fluid cells. The Lancet 1978-2 : 910-911.

Kan, Y. W., and A. M. Dozy (1980). Evolution of the hemoglobin S and C genes in world populations. Science 209 : 388-391.

Kanehisa, M. I. (1982). Los Alamos sequence analysis package for nucleic acids and proteins. Nucleic Acids Res. 10 : 183-196.

Kanehisa, M. I., and W. B. Goad (1982). Pattern recognition in nucleic acid sequences. II. An efficient method for finding locally stable secondary structures. Nucleic Acids Res. 10 : 265-277.

Kaplan, N., and C. H. Langley (1979). A new estimate of sequence divergence of mitochondrial DNA using restriction endonuclease mappings. J. Mol. Evol. 13 : 295-304.

Kaplan, N., and K. Risko (1981). An improved method for estimating sequence divergence of DNA using restriction endonuclease mappings. J. Mol. Evol. 17 : 156-162.

Kaplan, N., and K. Risko (1982). A method for estimating rates of nucleotide substitution using DNA sequence data. Theoret. Pop. Biol. 21 : 318-328.

Karlin, S., and A. Piazza (1981). Statistical methods for assessing linkage disequilibrium at the HLA-A,B,C loci. Ann. Hum. Genet. 45 : 79-94.

Karlin, S., and H. M. Taylor (1975). A First Course in Stochastic Processes. Second Edition, Academic Press, New York.

Karp, R. M. (1972). Reducibility among combinatorial problems. IN Complexity of Computer Computations. Miller, R. E., and J. W. Thatcher (Eds.), Plenum Press, New York, pp. 85-103.

Kashyap, R. L., and S. Subas (1974). Statistical estimation of parameters in a phylogenetic tree using a dynamic model of the substitutional process. J. Theor. Biol. 47 : 75-101.

Kimura, M. (1969). The number of heterozygous nucleotide sites maintained in a finite population due to steady flux of mutation. Genetics 61 : 893-903.

Kimura, M. (1980). A simple method for estimating evolutionary rates of base substitutions through comparative studies of nucleotide sequences. J. Mol. Evol. 16 : 111-120.

Kimura, M. (1981). Estimation of evolutionary distances between homologous nucleotide sequences. Proc. Natl. Acad. Sci. USA 78 : 454-458.

Kimura, M., and J. F. Crow (1964). The number of alleles that can be maintained in a finite population. Genetics 49 : 725-738.

Konkel, D. A., J. V. Maizel and P. Leder (1979). The evolution and sequence comparison of two recently diverged mouse chromosomal β-globin genes. Cell 18 : 865-873.

Korn, L. J., C. L. Queen and M. N. Wegman (1977). Computer analysis of nucleic acid regulatory sequences. Proc. Natl. Acad. Sci. USA 74 : 4401-4405.

Kurnit, D. M., and H. Hoehn (1979). Prenatal diagnosis of human genome variation. Ann. Rev. Genet. 13 : 235-258.

Lange, K., and M. Boehnke (1982). How many polymorphic marker genes will it take to span the human genome? Am. J. Human Genet. : 842-845.

Langley, C. H., E. A. Montgomery and W. F. Quattlebaum (1982). Restriction map variation in the Adh region of *Drosophila*. Proc. Natl. Acad. Sci. USA 79 : 5631-5635.

Lazowska, J., C. Jacq and P. P. Slonimski (1980). Sequence of introns and flanking exons in wildtype and *box3* mutants of cytochrome b reveals an interlaced splicing protein coded by an intron. Cell 22 : 333-348.

Li, W-H. (1975). Distribution of nucleotide differences between two randomly chosen cistrons in a finite population. Genetics 85 : 331-337.

Li, W-H. (1981). A simulation study of Nei and Li's model for estimating DNA divergence from restriction enzyme maps. J. Mol. Evol. 17 : 251-255.

Mather, K., and J. L. Jinks (1971). Biometrical Genetics. Second Edition, Chapman and Hall, London.

Maxam, A., and W. Gilbert (1977). A new method for sequencing DNA. Proc. Natl. Acad. Sci. USA 74 : 560-564.

McCallum, D., and M. Smith (1977). Computer processing of DNA sequence data. J. Mol. Biol. 116 : 29-30.

Messing, J., R. Crea and P. H. Seeburg (1981). A system for shotgun DNA sequencing. Nucleic Acids Res. 9 : 309-321.

Miklos, G. L. G., and A. C. Gill (1981). The DNA sequences of cloned complex satellite DNAs from Hawaiian *Drosophila* and their bearing on satellite DNA sequence conservation. Chromosoma 82 : 409-427.

Minsky, M., and S. Papert (1969). Learning. IN Perceptions. MIT Press, Cambridge, Massachusetts, pp. 161-187.

Montgomery, D. L., D. W. Leung, M. Smith, P. Shalit, G. Faye and B. D. Hall (1980). Isolation and sequence of the gene for iso-2-cytochrome c in *Saccharomyces cerevisiae*. Proc. Natl. Acad. Sci. USA 77 : 541-545.

Moore, G. W., J. Barnabas and M. Goodman (1973). A method for constructing maximum parsimony ancestral sequences on a given network. J. Theoret. Biol. 38 : 459-485.

Morrow, J. F. (1979). Recombinant DNA techniques. IN Methods in Enzymology 68: Recombinant DNA. Wu, R. (Ed.), Academic Press, New York, pp. 3-24.

Morton, N. E. (1955). Sequential tests for the detection of linkage. Am. J. Hum. Genet. 7 : 277-318.

Naylor, B. G. (1980). Radiation of the Amphibia Caudata: are we looking too far into the past? Evolutionary Theory 5 : 119-126.

Needleman, S. B., and C. D. Wunsch (1970). A general method applicable to the search for similarities in the amino acid sequence of two proteins. J. Mol. Biol. 48 : 443-453.

Nei, M. (1975). Molecular Population Genetics and Evolution. North Holland, Amsterdam and Oxford.

Nei, M., and W-H. Li (1979). Mathematical model for studying genetic variation in terms of restriction endonucleases. Proc. Natl. Acad. Sci. USA 76 : 5269-5273.

Nei, J., and F. Tajima (1981). DNA polymorphism detectable by restriction endonucleases. Genetics 97 : 145-163.

Neyman, J. (1971). Molecular studies of evolution: a source of novel statistical problems. IN Statistical Decision Theory and Related Topics. Gupta, S. S., and J. Yackel (Eds.), Academic Press, New York, pp. 1-27.

Nussinov, R., and A. B. Jacobson (1980). Fast algorithm for predicting the secondary structure of single-stranded RNA. Proc. Natl. Acad. Sci. USA 77 : 6309-6313.

Orcutt, B. C., D. G. George, J. A. Frederickson and M. O. Dayhoff (1982). Nucleic acid sequence database computer system. Nucleic Acids Res. 10 : 157-174.

Parker, R. C., R. M. Watson and J. Vinograd (1977). Mapping of closed circular DNAs by cleavage with restriction endonucleases and calibration by agarose gel electrophoresis. Proc. Natl. Acad. Sci. USA 74 : 851-855.

Pauling, L., and E. Zuckerkandl (1963). Chemical paleogenetics. Molecular "restoration studies" of extinct forms of life. Acta Chem. Scand. 17 : S9-S16.

Pavlakis, G. N., B. R. Jordan, R. M. Wurst and J. N. Vournakis (1979). Sequence and secondary structure of *Drosophila melanogaster* 5.8S and 2S rRNAs and of the processing site between them. Nucleic Acids Res. 7 : 2213-2238.

Pavlakis, G. N., R. E. Lockard, N. Vamvakopoulos, L. Rieser, U. L. RajBhandary and J. N. Vournakis (1980). Secondary structure of mouse and rabbit α- and β-glogin mRNAs: differential accessibility of α and β initiator AUG codons towards nucleases. Cell 19 : 91-102.

Peacock, W. J., E. S. Dennis, M. M. Rhoades and A. J. Pryor (1981). Highly repeated DNA sequence limited to knob heterochromatin in maize. Proc. Natl. Acad. Sci. USA 78 : 4490-4494.

Pearson, W. R. (1982). Automatic construction of restriction site maps. Nucleic Acids Res. 10 : 217-227.

Penny, D. (1982a). Towards a basis for classification: the incompleteness of distance measures, incompatibility and phenetic classification. J. Theor. Biol. 96 : 129-142.

Penny, D. (1982b). Branch and bound methods for minimal trees and maximal cliques. Syst. Zool. (in press).

Petes, T. D., and D. Botstein (1977). Simple Mendelian inheritance of the reiterated ribosomal DNA of yeast. Proc. Natl. Acad. Sci. USA 74 : 5091-5095.

Pielou, E. C. (1977). Mathematical Ecology. John Wiley, New York.

Pipas, J. M., and J. E. McMahon (1975). Method for predicting RNA secondary structure. Proc. Natl. Acad. Sci. USA 72 : 2017-2021.

Queen, C. L., and L. J. Korn (1980). Computer analysis of nucleic acids and proteins. IN Methods in Enzymology. Grossman, L., and K. Moldave (Eds.), Vol. 65, Academic Press, New York, pp. 595-609.

Queen, C., M. N. Wegman and L. J. Korn (1982). Improvement to a program for DNA analysis: a procedure to find homologies among many sequences. Nucleic Acids Res. 10 : 449-456.

Rao, C. R. (1965). Linear Statistical Inference and Its Applications. John Wiley, New York.

Renwick, J. H. (1971). The mapping of human chromosomes. Ann. Rev. Genet. 5 : 81-120.

Robbins, H. E. (1944). On the measure of a random set. Ann. Math. Stat. 15 : 70-74.

Robbins, H. E. (1945). On the measure of a random set. II. Ann. Math. Stat. 16 : 342-347.

Roberts, R. J. (1982). Restriction and modification enzymes and their recognition sequences. Nucleic Acids Res. 10 : r117-r144.

Rosenblatt, F. (1959). Two theorems of statistical separability in the Perceptron. IN Mechanization of Thought Processes. HMSO, London, pp. 421-456.

Rotwein, P., R. Chyn, J. Chirgwin, B. Cordell, H. M. Goodman and M. A. Permutt (1981). Polymorphism in the 5' - flanking region of the human insulin gene and its possible relation to type 2 diabetes. Science 213 : 1117-1120.

Salser, W. (1977). Globin mRNA sequences: analysis of base pairing and evolutionary implications. Cold Spring Harbor Symp. Quant. Biol. 42 : 985-1002.

Sanger, F., S. Nicklen and A. R. Coulson (1977). DNA sequencing with chain-terminating inhibitors. Proc. Natl. Acad. Sci. USA 74 : 5463-5467.

Sankoff, D. (1975). Minimal mutation trees of sequences. SIAM J. Appl. Math. 28 : 35-42.

Sankoff, D. D., and P. Rousseau (1975). Locating the vertices of a Steiner tree in arbitrary metric space. Math. Programming 9 : 240-246.

Schaffer, H. E., and R. R. Sederoff (1981). Improved estimation of DNA fragment lengths from agarose gels. Anal. Biochem. 115 : 113-123.

Schleif, R., and J. Hirsh (1980). Electron microscopy of proteins bound to DNA. IN Methods in Enzymology 65 (1): Nucleic Acids. Grossman, L., and K. Moldave (Eds.), Academic Press, New York, pp. 885-896.

Schneider, T. D., G. D. Stormo, J. S. Haemer and L. Gold (1982). A design for computer nucleic acid sequence storage, retrieval, and manipulation. Nucleic Acids Res. 10 : 3013-3024.

Schroeder, J. L., and F. R. Blattner (1978). Least-squares method for restriction mapping. Gene 4 : 167-174.

Schroeder, J. L., and F. R. Blattner (1982). Formal description of a DNA oriented computer language. Nucleic Acids Res. 10 : 69-84.

Sege, R., D. Soll, R. H. Ruddle and C. Queen (1981). A conversational system for the computer analysis of nucleic acid sequences. Nucleic Acids Res. 9 : 437-444.

Sellers, P. H. (1974). On the theory and computation of evolutionary distances. SIAM J. Appl. Math. 26 : 787-793.

Sellers, P. H. (1979). Pattern recognition in genetic sequences. Proc. Natl. Acad. Sci. USA 76 : 3041.

Shah, D. M., and C. H. Langley (1979). Inter- and intraspecific variation in restriction maps of *Drosophila* mitochondrial DNAs. Nature 281 : 696-699.

Shapiro, R., and J. Hachmann (1966). The reaction of guanine derivatives with 1,2-dicarbonyl compounds. Biochemistry 5 : 2799-2807.

Shen, S., J. L. Slightom and O. Smithies (1981). A history of the human fetal globin gene duplication. Cell 26 : 191-203.

Shepherd, J. C. W. (1981). Method to determine the reading frame of a protein from the purine/pyrimidine genome sequence and its possible evolutionary justification. Proc. Natl. Acad. Sci. USA 78 : 1596-1600.

Shine, J., and L. Dalgarno (1974). The 3'-terminal sequence of *Escherichia coli* 16S ribosomal RNA: complementarity to nonsense triplets and ribosome binding sites. Proc. Natl. Acad. Sci. USA 71 : 1342-1346.

Shulman, M. J., C. M. Steinberg and N. Westmoreland (1981). The coding function of nucleotide sequences can be discerned by statistical analysis. J. Theor. Biol. 88 : 409-420.

Siegel, S. (1956). Nonparametric Statistics for the Behavioral Sciences. McGraw-Hill, New York.

Skolnick, M., and U. Franke (1982). Report of the Committee on Human Gene Mapping by Recombinant DNA Techniques. VI*th* International Workshop on Human Gene Mapping, Oslo, June 29 - July 3, 1981.

Solomon, H. (1978). Geometric Probability. Soc. Ind. Appl. Math., Philadelphia.

Southern, E. M. (1975). Detection of specific sequences among DNA fragments separated by gel electrophoresis. J. Mol. Biol. 98 : 503-517.

Southern, E. M. (1979a). Gel electrophoresis of restriction fragments. IN Methods in Enzymology 68 : Recombinant DNA. Wu, R. (Ed.), Academic Press, New York, pp. 152-176.

Southern, E. M. (1979b). A preparative gel electrophoresis apparatus for large scale separations. Anal. Biochem. 100 : 304-318.

Southern, E. M. (1979c). Measurement of DNA length by gel electrophoresis. Anal. Biochem. 100 : 319-323.

Staden, R. (1977). Sequence data handling by computer. Nucleic Acids Res. 4 : 4037-4051.

Staden, R. (1978). Further procedures for sequence analysis by computer. Nucleic Acids Res. 5 : 1013-1015.

Staden, R. (1979). A strategy of DNA sequencing employing computer programs. Nucleic Acids Res. 6 : 2601-2610.

Staden, R. (1980a). A computer program to search for tRNA genes. Nucleic Acids Res. 8 : 817-825.

Staden, R. (1980b). A new computer method for the storage and manipulation of DNA gel reading data. Nucleic Acids Res. 8 : 3673-3694.

Staden, R., and A. D. McLachlan (1982). Codon preference and its use in identifying protein coding regions in long DNA sequences. Nucleic Acids Res. 10 : 141-156.

Staehelin, M. (1959). Inactivation of virus nucleic acid with glyoxal derivatives. Biochim. Biophys. Acta 31 : 448-454.

Stefik, M. (1978). Inferring DNA structures from segmentation data. Artificial Intelligence 11 : 85-114.

Stevens, W. L. (1939). Solution to a geometrical problem in probability. Ann. Eugenics 9 : 315-320.

Stormo, G. D., T. D. Schneider and L. M. Gold (1982). Characterization of translational initiation sites in *E. coli*. Nucleic Acids Res. 10 : 2971-2996.

Studnika, G. M., G. M. Rahn, I. W. Cummings and W. A. Salser (1978). Computer method for predicting the secondary structure of single-stranded RNA. Nucleic Acids Res. 5 : 3365-3387.

Sutcliffe, J. G. (1978). Complete nucleotide sequence of the *Echerichia coli* plasmid pBR322. IN Cold Spring Harbor Symposia on Quantitative Biology 43 : 77-90.

Swofford, D. L. (1981). On the utility of the distance Wagner procedure. IN Advances in Cladistics. Funk, V. A., and D. R. Brooks (Eds.), New York Botanical Garden, New York, pp. 25-43.

Takahata, N., and M. Kimura (1981). A model of evolutionary base substitutions and its application with special reference to rapid change of pseudogenes. Genetics 98 : 641-657.

Templeton, A. R. (1983). Phylogenetic inference from restriction endonuclease cleavage site maps with particular reference to the evolution of man and the apes. Evolution (in press).

Templeton, A. R., R. DeSalle and V. Walbot (1982). Speciation and inferences on rates of molecular evolution from genetic distances. Heredity 47 : 439-442.

Thompson, E. A., K. Kravitz, J. Hill and M. H. Skolnick (1978). Linkage and the power of a pedigree structure. IN Genetic Epidemiology. Morton, N. E., and C. S. Chung (Eds.), Academic Press, New York, pp. 247-253.

Thomson, G. (1977). The effect of a selected locus on linked neutral loci. Genetics 85 : 753-788.

Tinoco, I., O. C. Uhlenbeck and M. D. Levine (1971). Estimation of secondary structure in ribonucleic acids. Nature 230 : 362-367.

Tolstoshev, C. M., and R. W. Blakesley (1982). RSITE: a computer program to predict the recognition sequence of a restriction enzyme. Nucleic Acids Res. 10 : 1-17.

Upholt, W. B. (1977). Estimation of DNA sequence divergence from comparison of restriction endonuclease digests. Nucleic Acids Res. 4 : 1257-1265.

Upholt, W. B., and I. B. Dawid (1977). Mapping of mitochondrial DNA of individual sheep and goats: rapid evolution in D loop region. Cell 11 : 571-583.

Vasilenko, S., and G. T. Babkina (1965). Isolation and properties of the ribonuclease from cobra venom. Biokhimiya 30 : 705-712. (English translation: Biochemistry USSR 30 : 606-611.)

Vogel, F., and M. Kopun (1977). Higher frequencies of transitions among point mutations. J. Mol. Evol. 9 : 159-180.

Waterman, M. C., T. F. Smith and W. A. Beyer (1976). Some biological sequence metrics. Adv. Math. 20 : 367-387.

Watterson, G. A. (1975). On the number of segregating sites in genetic models without recombination. Theor. Pop. Biol. 7 : 256-276.

Wilson, A. C., S. S. Carlson and T. J. White (1977). Biochemical evolution. Ann. Rev. Biochem. 46 : 573-639.

Wollenzien, P., J. E. Hearst, P. Thammana and C. R. Cantor (1979). Base-pairing between distant regions of the *Escherichia coli* 16S ribosomal RNA in solution. J. Mol. Biol. 135 : 255-269.

Wu, R. (Ed.) (1979). Methods in Enzymology 68 : Recombinant DNA. Academic Press, New York.

Wu, R., R. Bambara and E. Jay (1974). Recent advances in DNA sequence analysis. IN CRC Critical Reviews in Biochemistry 2 : 455-512.

Wyman, A. R., and R. White (1980). A highly polymorphic locus in human DNA. Proc. Natl. Acad. Sci. USA 77 : 6754-6758.

Yang, R. C., J. Lis and R. Wu (1979). Elution of DNA from agarose gels after electrophoresis. IN Methods in Enzymology 68 : Recombinant DNA. Wu, R. (Ed.), Academic Press, New York, pp. 176-182.

Zain, S., T. R. Gingeras, P. Bullock, G. Wong and R. E. Gelinas (1979). Determination and analysis of adenovirus-2 DNA sequences which may include signals for late messenger RNA processing. J. Mol. Biol. 135 : 413-433.

Zimmer, E. A. (1980). Evolution of Globin Genes. Ph.D. Dissertation, University of California, Berkeley.

Zouros, E. (1979). Mutation rates, population sizes and amounts of electrophoretic variation of enzyme loci in natural populations. Genetics 92 : 623-646.

Zuker, M., and P. Stiegler (1981). Optimal computer folding of large RNA sequences using thermodynamics and auxiliary information. Nucleic Acids Res. 9 : 133-148.

INDEX

I

K

L

M